Springer Praxis Books

More information about this series at http://www.springer.com/series/4097

Michael Carroll and Rosaly Lopes

Antarctica: Earth's Own Ice World

 **Springer**

Published in association with
Praxis Publishing
Chichester, UK

Michael Carroll
Littleton, Colorado, USA

Rosaly Lopes
Jet Propulsion Laboratory
Pasadena, California, USA

SPRINGER PRAXIS BOOKS IN POPULAR SCIENCE

Springer Praxis Books
ISBN 978-3-319-74623-4   ISBN 978-3-319-74624-1   (eBook)
https://doi.org/10.1007/978-3-319-74624-1

Library of Congress Control Number: 2018933382

Cover photo by the authors. In the background rises the mist-shrouded summit of Mount Erebus.

Printed on acid-free paper

This Springer imprint is published by the registered company Springer International Publishing AG part of Springer Nature.
The registered company address is: Gewerbestrasse 11, 6330 Cham, Switzerland

Contents

Acknowledgements

Our thanks go first and foremost to the NSF's Peter West and Valentine Kass, who manage the Artists and Writers Program, and to Michael Lucibella, who worked out our logistics for the field. A special commendation goes to Elaine Hood, our McMurdo "tour guide" and implementer extraordinaire. These remarkable people were the vanguard of all the personnel in the USAP who made our trip possible and successful. Our mountaineer Evan Miller showed us the ropes—literally—making our stay at Fang Glacier tolerable and our voyage across Erebus inspiring and safe. Jani Radebaugh, Chris McKay, and Nick Baggarly served as our polar fashion advisors (did we look good? Isn't that what counts?). Thanks to Glen Nagle of Canberra, Australia, for the beautiful Enceladus mosaic via Astro0. NASA, JPL, and the Space Science Institute are always of supreme help with imagery of the cosmos. Rosaly thanks her late cat Blanca, who was as white as snow and as fierce as Erebus, for initial inspiration to write about Antarctica. Mike thanks his long-suffering wife Caroline for support of this expedition and Rosaly's son Tommy Gautier for the use of his spiffy SLR. Last but by no means least, deep thanks to our editor at Springer, Maury Solomon, who supported this project from the very start and saw it through to completion.

Preface: Beginnings of an Adventure

Fig. A The simmering cauldron atop Mount Erebus drapes icy mists across a royal blue Antarctic sky (photo by the authors).

Rising 3795 m (12,448 feet) above the glistening plain of Antarctica's Ross Sea Ice Shelf, Mount Erebus is the southernmost active volcano on Earth. First seen by James Clark Ross' expedition of 1841, the volcano offers insights into discoveries never dreamed of by those early explorers, revelations that extend across the moons and planets of our Solar System. The mountain's resemblance to other worlds draws researchers from diverse fields, and in 2016, it drew us on a 30,000-km voyage to its flanks. To the sponsoring agency, the National Science Foundation, seasoned researcher Rosaly Lopes was a must: she is a planetary geologist at the Jet Propulsion Laboratory in Pasadena, California, an expert on volcanoes of the outer Solar System. Space artist and science journalist Michael Carroll's appeal to the NSF lay in the fact that he has often depicted volcanic landscapes on other worlds and written about them as well. Together we travelled to the Harsh Continent under the auspices of the National Science Foundation's *Artists and Writers program*, part of the United States Antarctic Program. Our trip was made possible by the networks and resources laid down across the southern regions over more than a century of exploration and discovery.

Michael Carroll   Rosaly Lopes
Littleton, CO, USA   Pasadena, CA, USA

1

The Lure of the Poles

Fig. 1.1. Mount Erebus rises some 3,795 meters (12,500 feet) above the Ross Ice Shelf. Fuel depots line a roadway emblazoned across the snow and ice, while flags mark dangerous sites or routes back to safety in a whiteout (Photo by Michael Carroll)

M. Carroll, R. Lopes, *Antarctica: Earth's Own Ice World*, Springer Praxis Books, https://doi.org/10.1007/978-3-319-74624-1_1

Goethe said, "The ideal of beauty is simplicity and tranquility." He never saw Mount Erebus. The 3,795 meter (12,448 foot) tall volcano soars majestically into Antarctic skies, collared by mist and fog. Multiple craters crown its summit, while its fumaroles and vents build complex towers and ice columns. Erebus eruptions generate the rarest of volcanic crystals, seen on only one other volcano on Earth. And while the mountain may seem tranquil most of the time, its beauty has come at the hand of violent forces.

Erebus is the southernmost active volcano in the world, and it is very active. Meter-sized lava bombs occasionally soar over the rim, impacting its flanks with molten rock. In fact, a science outpost called the Upper Erebus Hut had to be abandoned when two such projectiles landed just beyond the structure. Erebus presents a hostile environment to explorers, but one with unique formations that may offer some of the best terrestrial insights into the features on the icy moons of the outer Solar System and, in particular, Saturn's dynamic moon Enceladus. With Enceladus in mind, the authors Lopes and Carroll embarked on their 2017 expedition to Antarctica. Our trip added to a rich litany of explorations completed over several centuries.

Legends Before History

Writings of the ancient Greeks mention a theoretical southern continent as early as 350 BC. The Greeks reasoned that since the northern part of the world, the arctic, was watched over by the constellation of the bear (Arktos), it made sense that the world's equilibrium would demand a similar cold region in the south. They coined the term "Ant-Arktos" for the region opposite the lands watched over by the bear. Greek astronomer Ptolemy suggested the existence of an "unknown southern land" that would balance out the continents farther north in the known world. Yet for the Greeks, Antarctica did not extend beyond theory (it is worth noting that Greek sailor Ptheas may have reached as far north as Iceland, making him perhaps the first polar adventurer). Terra Australis, the great southern continent, appears in legends as early as 650 AD, centuries before medieval Europeans theorized its existence. The Maori, indigenous people of New Zealand[1], tell of a great war canoe that ventured to the southern sea ice. Captained by the mariner Ui-te-Rangiora, the small fleet of reed boats may have made it far enough south to see the Ross Ice Shelf. The legend refers to the southern ocean as Tai-uka-a-pia, Maori for "sea foaming like arrowroot" (when scraped for cooking, arrowroot is the consistency of snow). The ancient text says, "These were those wonderful things: the rocks that grow out of the sea, in the region beyond Rapa."[2] The rocks growing from the sea may well be a phrase signifying icebergs and floes.

Even into the Middle Ages, legends spoke of a vast continent spreading across the South Pole. 16th century exploration by sea pushed farther south than before. In 1520, Magellan suggested that the territory south of the Strait of Magellan – Tierra del

---

[1] The Maori, peoples migrating from eastern Polynesia, populated New Zealand in several waves between 1250 and 1300 AD.

[2] As related in the text *Hawaiki: the Original Home of the Maori*, by S. Percy Smith (Cambridge University Press 1910)

Fuego – could be the northern edge of the long-theorized vast southern continent (it is not in fact a solid landmass, but rather a series of islands at the tip of South America). In 1578, Sir Francis Drake sailed his ship, the famed *Golden Hind*, through the same waters. A storm blew him southeast around Cape Horn. On this journey, he observed the true nature of Tierra del Fuego and declared that any southern continent must be much farther south.

Yet more stormy weather contributed to southern exploration in 1619, when Spanish sailors Bartoleme and Gonzalo Garcia de Nodal found themselves within the chain of islands called Islas Diego Ramirez. Their unplanned journey marked the farthest south venture for the next 156 years. Another tempest carried the English trader Anthony de la Roche to the 55° south line, where he found refuge in a bay on what was probably South Georgia Island. On the trip, he sighted what he thought was the coastline of the southern continent. His conclusion was undoubtedly influenced by maps published at the time by the Dutch East India Company, which delineated the shoreline of the imaginary "Terra Australis Incognita."

The true extent of any mythical continent was constrained by the voyage of the British explorer Captain James Cook, who circumnavigated the globe in the high southern latitudes in 1772-1775. At one point, Cook reached a southern latitude of 71° before being blocked by the ice pack. Cook's route demonstrated that if a southern continent existed, it must have been south of the 60° parallel.

The First Breakthrough

Dante's medieval literary masterpiece *The Inferno* describes Hell as a place arranged in concentric circles. The heart of the dark realm is accessible only through various outer boundaries, each more treacherous to cross than the last. At Hell's center, Satan holds court within a lake of eternally frozen ice. To early explorers, reaching the heart of Antarctica posed a similar challenge. The earliest adventurers reached only the outer fringes of the sea ice, a seemingly impenetrable wall blocking the way south. Beyond the sea ice, others caught glimpses of frozen walls and icebergs rearing up, the solid cliffs of the permanent sea ice sheet. Beyond it, the mountains of the Antarctic continent rose, at first viewed only at a distance but then seen progressively closer as explorers ventured into the interior. The revelations of Antarctica's true nature initially came in starts and stops. Travelers reported northward-drifting icebergs with rocks sticking out of their flanks, but had they come from solid land or an archipelago of Antarctic islands? As travel became more reliable, the push into the mysterious frozen lands became more sustained and steadier. Still, the coastal explorations and first tentative outposts would not give way to more permanent bases for many decades, and the remote camps of the interior had to wait even longer.

Half a century after Cook, sea-going pioneers finally glimpsed the shoreline of Antarctica, although which explorer was first is still a matter of contention. The Russian voyager Fabian Gottleib von Bellingshausen is one of three candidates. Bellingshausen was an admiral in the Russian navy. While commanding the ships *Vostok* and *Mirny* in 1820, the admiral's crew sighted the islands of Peter I and Alexander I. This may have been the first solid land seen within the Antarctic Circle.

Fig. 1.2. Maori totara wood sculpture *Pou Whenua* ("Navigator of the heavens"), dedicated by the Maori people to New Zealand's distinctively green Scott Base in 2013. The event commemorated Scott Base's 56th anniversary (Photo by Michael Carroll)

Another early sighting took place on November 17, 1820. At the time, several sealers, most notably the Englishman James Wedell, ventured into Antarctic waters in search of new hunting grounds. Wedell made it as far as 74°S, a record that was to stand for 80 years (although it is possible that the American sea captain Benjamin Morell may have made it that far just a month later). American sealer Nathaniel Brown Palmer made landfall in Antarctica while in search of seal rookeries. Remarkably, he made the voyage in a small, 14-m (47-foot) long sloop named *Hero*. The able seaman called the area "Palmer Land." Today, the coast and several islands on this part of Western Antarctica, bordering the Orleans Strait, bear his name. Palmer went on to become an instrumental designer of a new class of sailing vessel known as the clipper ship.[3]

At the same time, British Royal Naval officer Edward Bransfield was commissioned to sail the two-masted brig *Williams* south from Chile to survey the newly discovered South Shetland Islands, scattered very near the Antarctic Peninsula. Making his way farther to the southwest, past Deception Island, Bransfield finally spotted the Trinity Peninsula. Its Prime Head is the northernmost limit of the Antarctic continent.[4] He recorded "high mountains covered with snow." What he saw were the mainland peaks now known as Mt. Bransfield and Mt. Jacquinot. Historians have studied and compared the journals of both Bransfield and von Bellingshausen, and most credit the latter with the actual discovery of Antarctica, as it appears that von Bellingshausen arrived two days before Bransfield.

As Europe recovered from war and unrest at the opening of the 19th century, Antarctica welcomed a host of European explorers. One of the most influential was James Clark Ross, who led two expeditions to the Harsh Continent. Ross set out on the *HMS Erebus* and the *HMS Terror* (commanded by his fellow seaman Francis Crozier) in 1839. His choice of ships was a wise one: the two boats were built with reinforced hulls to withstand the stresses of the mortars they had been designed to carry (the ships were known as bomb-vessels). Their tough exteriors made them the ideal option for a voyage through polar ices. The expedition was to set up magnetic observatories at several sites to monitor the Earth's magnetic fields, ultimately establishing a permanent station in Tasmania. Ross's team carried out measurements at Saint Helena Island and in the Kerguelen islands at the Cape of Good Hope. Only two days after departing from the Kerguelens, a hurricane separated the ships. They were reunited at Hobart, Tasmania on August 16, 1840. From there, Ross was determined to go south to discover where the Earth's magnetic South Pole was centered after learning that both the American Charles Wilkes and the French explorer Dumont d'Urville had been doing magnetic field research in the area. Wilkes actually gifted his charts to Ross to aid his studies.

Ross mapped hundreds of miles of Antarctic coastline. The Ross Ice Shelf is named after him (although he had named it the Victoria Barrier), as is Ross Island, home of the active volcano Mount Erebus – named after Ross's ship. The members of the Ross 1841 expedition were the first to document an eruption of Mount Erebus, which they witnessed from afar. Erebus simmers next to an extinct volcano christened after Ross's second ship, the *Terror*. Their voyage was one of discovery and science. Aboard ship with Ross was Joseph Hooker, who would later become president of the Royal Society and a close colleague of the naturalist Charles Darwin. Hooker took notes on the Erebus eruption, describing "dazzling beautiful peaks of snow which, when the sun approached the horizon, reflected the most brilliant tints of golden yellow and scarlet; and then to see the dark

---

[3] Palmer could never have imagined that some two centuries later, a mission called the Europa Clipper will head to the outer Solar System.

[4] Just 20 km to the south lies Esperanza Base, one of Argentina's year-round Antarctic outposts.

Fig. 1.3. "Erebus and Terror in New Zealand, August 1841," painted by John Wilson Carmichael (Oil on canvas ca. 1847)

cloud of smoke, tinged with flame, rising from the volcano in a perfectly unbroken column, one side jet-black, the other giving back the colors of the sun..."

In 1838, at about the time of Ross' trailblazing expedition, the US Congress funded an Antarctic trip led by Charles Wilkes. Antarctic exploration had fired the American imagination, and the government felt the time was ripe for a new, more well-funded and complex American mission. Congress' mandate was to aid commerce and navigation, but they included a clause about extending information on unknown territory and promoting general knowledge. The *Wilkes Expedition* departed in 1838 and reconnoitered territory from Brazil to Chile and Tierra del Fuego, even venturing into Australia, New Zealand, the Philippines and the East Indies. Wilkes pushed down into Antarctica on two occasions, in 1839 and 1840, during the austral summers. He followed the edge of the southern ice pack for a 2,400-kilometer stretch, spotting the Antarctic landmass at a distance on several different occasions and proving the existence of a solid continent. His team returned with surveys of geology, zoology, and even anthropological specimens from some of the islands.

Over 30 years later, 1874 brought the illustrious voyage of the *H.M.S. Challenger*. In February of that year, the *Challenger* became the first steam-powered ship to enter Antarctic waters. Her primary power still came from her massive sails, some 1,480 square meters (16,000 square feet) under full sail. *Challenger* was a science vessel. The Royal Navy's converted warship was tasked with carrying out the first worldwide oceanographic research mission, a voyage that was to cover 68,900 miles of ocean. The ship was outfitted with laboratories and a darkroom for developing photographs. It is thought that the *Challenger Expedition* was the first to make extensive use of photography as a research tool. Engineers also outfitted the ship with a steam-powered dredging system for bringing up samples from deep water and the seabed. A globe with a hole on the side could also be lowered to the sea floor, dragged, and reeled back in for sampling.

On February 24, 1874, *Challenger* nearly met with disaster. In the early morning hours, the crew was in process of dredging—an operation often met with curiosity—when a gale-force blizzard arrived. Pulling the dredge up as quickly as possible, the ship used steam and sail to head for shelter on the leeward side of a large iceberg. But the stormy whitecaps threw the ship against the iceberg, tearing away the jib boom. The crew fired up all four of the ship's boilers and steamed away from the berg, dragging the jib behind them. Although it was recovered, the ship now faced low visibility and rough seas. At 3 pm, another iceberg loomed dangerously close. The ship had to ply full steam astern with all hands on deck, but as the weather calmed, the ship made it through the bergs. Three days later, with a battery of experiments, surveys and investigations complete, the ship again raised all sails and headed northeast toward open sea and civilization in Melbourne, Australia.

Among the *Challenger's* 243-strong crew of engineers, scientists, officers, and seamen was John James Wild, an oceanographer and artist. Wild was one of a growing number of scientists who documented their research in the form of meticulously detailed art. After the four-year voyage, Wild wrote the book *At Anchor, a Narrative of Experiences Afloat and Ashore During the Voyage of H.M.S.* Challenger *from 1872 to 1876*. In conjunction with the *Challenger's* expedition, he wrote the paper *Thalassa, An Essay on the Depth, Temperature and Currents of the Ocean*. Wild was Swiss, and for his work, he received an honorary doctorate degree from the University of Zurich. Wild's presence on the *Challenger* team prophesied the spirit and intent of the NSF's artists and writers program over a century later.

Building on the rich return of specimens by Wilkes, the *Challenger* crew and others, the 1892 Norwegian expedition of Carl Larsen landed on the Seymore Island just off the Antarctic Peninsula, where they unearthed a host of fossils. These specimens became the first evidence that Antarctica may have enjoyed a warmer climate in its past. Larsen went on to build the first whaling station in Antarctica on South Georgia Island. Swedish and German researchers soon followed. In 1895, Norwegian whaler Henryk Johan Bull led a two-year expedition that included a landing at Cape Adare, Antarctica. On board was pioneer explorer Carsten Borchgrevink, who was first to collect plant life in Antarctica in the form of lichens. Up until his discovery, the scientific consensus was that plants could not survive so far south.

By 1902, the stage was set for the Heroic Age of Exploration. Robert Falcon Scott set out from what is now known as the McMurdo Sound, bound for his first attempt at the South Pole, with Edward Wilson and Ernest Shackleton in tow. Suffering from malnutrition and snow blindness, the brave trio had to turn back at 82°17′ S. They survived the trek home, completing an astounding round trip of nearly 5,000 kilometers on foot.

In the late 1800s, Mount Erebus, our ultimate goal, remained a distant pinnacle of snowy mystery. The first summiting of Mount Erebus would not take place until 1908, when five members of Sir Ernest Shackleton's *Nimrod Expedition* spent nearly a week ascending the mountain, braving a blizzard and -30°F temperatures (see below). Four years later, members of Robert Scott's *Terra Nova Expedition* surveyed Erebus and took geological samples of its unique "Erebus crystals." Remnants of two of their campsites – the Upper Summit Camp and the Lower Camp E – are now internationally recognized historical sites.

The "Heroic Age"[5] of Exploration: Northern Exposures

In the 19th and 20th centuries, the poles themselves called to explorers: there were fresh vistas to be seen, hostile environments to conquer, new frontiers to explore. In particular, the poles had a compelling attraction to those with an explorer's wanderlust. The North Pole had another appeal: navigators yearned for a northwest passage, a shortcut over the pole from the Americas to Europe and Asia.

Norwegian adventurer Roald Amundsen, one of the great Antarctic pioneers, was first to cross the Northwest Passage, announcing his success by telegram from the tiny village of Eagle, Alaska. Although Amundsen had made it through, he found that the route was too shallow for practical shipping lanes.

Many rushed to reach the north and south geographic poles. First to the North Pole may have been Admiral Robert Peary, who claimed to reach the top of the world on April 6, 1909. Peary travelled using dogsleds and set up three successive support teams along the way. His claim is in dispute. Others – most notably Frederick Cook – also claimed to have reached the pole, and still others died trying. In 1825, William Edward Parry lost his ship, the *HMS Fury*, to the ice near Baffin Island. Two years later, Parry made it to 82°45' N, the farthest north at the time. The year 1871 saw a polar attempt by Charles Francis Hall's *Polaris Expedition*, sponsored by the US government. The expedition was hounded by poor leadership; the crew nearly mutinied on several occasions. The Polaris surpassed Parry's record but could not make it to the pole itself. As the Polaris crew progressed northward from a wintering site on the Greenland coast, Hall left the ship and went ahead on a sledging trip to survey. Upon his return, he fell critically ill and blamed his crew for "poisoning" him. Hall died, but no charges were filed against the crew. In 1968, his body was exhumed and an autopsy revealed high levels of arsenic.

The US again tried to reach the pole with the Navy's *Jeanette Expedition* (1879-1881), commanded by George DeLong. DeLong's route originated on the Pacific side, winding its way through the Bering Strait. But soon the good ship *Jeanette* found herself trapped in the ice. DeLong led his crew by boat and sled to the Lena River Delta in Siberia, but in the course of doing so, more than half of the crew – including their leader – died of cold or starvation before rescue. It took the *Jeanette* two years to sink some 300 miles north of the Siberian coast.

Famous Norwegian explorer Fridtjof Nansen attempted to reach the pole during a three-year voyage from 1893-1896 using a creative approach. The seafarer allowed his ship, the specially designed *Fram*, to become icebound in the eastern Arctic at the New Siberian Islands. From there, Nansen planned to let the trapped ship drift in the pack ice, with the Arctic's own sea ice carrying it to the pole. His inspiration was the wreckage of the ship *Jeanette*, which appeared to have drifted across the North Pole before being discovered on the southwest coast of Greenland. Nansen's ship drifted slowly for 18 months, and the Norwegian explorer became impatient. He set out with a colleague for the pole on foot

---

[5]The Heroic Age opened at the end of the 1800s and ended at the close of Shackleton's Trans-Antarctic expedition in 1917 when its survivors arrived in Wellington, New Zealand.

with the aid of a team of dogs. The two did not reach their destination, but they did survive the trip back from the farthest north point yet explored, at the latitude of 86° 13.6' N.

Roald Amundsen was part of a 1926 expedition to fly over the pole, an excursion that recorded its first undisputed sighting (Amundsen was also first to attain both poles). The flight was carried out by Italian aviator and engineer Umberto Nobile aboard his dirigible, the *Norge,*[6] which Amundsen had traveled to Rome to purchase. He was accompanied by his colleague Hjalmar Riiser-Larsen, founder of the Royal Norwegian Air Force. At the time, the American explorer Admiral Richard Byrd was drawing plans to make it to the pole as well, but Amundsen declared that his was a far larger vision. He wanted to do a survey of the entire region, referring to the North Pole as "merely a station on the way." The *Norge* drifted over the pole at 1:25 am on May 12, 1926.

Two years later, the French government backed Nobile in a second series of polar flights. In the spring of 1928, the airship *Italia* departed on the first of five planned flights. The *Italia* was essentially identical to the *Norge* in design, spanning a length of 145 meters (348 feet) with a diameter of 19 m (64 ft). It carried a crew of 20. During the first attempt, the airship was forced to return after eight hours due to weather. The second flight charted an impressive 4,000 kilometers of arctic wilderness under fine weather conditions. But it was the third flight that became the most famous.

May 23, 1928 saw the departure of the *Italia* along the coast of Greenland to the North Pole. The ship carried equipment with which to deploy ground crews for surface exploration. But two days out, the weather turned. Ice buildup and engineering problems caused the *Italia* to crash on the sea ice, tearing the gondola from the gas envelope of the dirigible. Six of Nobile's comrades were trapped in the envelope as it drifted away on the wind. Ten crew members were marooned in the wrecked gondola on the sea ice, one of whom died from the impact. The other nine erected a tent using silk from their supplies and sections of the ruined envelope. They dyed the tent red using the dye from glass altitude "bombs" used to measure the ship's altitude during flight. They managed to save a small radio and collected food rations, many of which had been tossed to them by Chief Flight Engineer Ettore Arduino. Upon seeing the men left behind on the ice, Arduino had tossed as many supplies as he could to them as the envelope carried him and the rest of the crew away in the arctic gale, undoubtedly saving the lives of the stranded crew. Those trapped in the envelope – including Adruino – were never seen again.

Five countries carried out aerial searches after a Soviet amateur radio operator heard the crew's SOS. Eventually, all but one of the men in the gondola were rescued in a series of flights and the arrival of the Soviet icebreaker ship *Krasin.*

The search and rescue operations for the crew of the *Italia* entailed one of the great ironies of polar exploration. After a lifetime of daring exploration, careful planning and successful triumphs to both poles, Roald Amundsen vanished while attempting to locate and rescue Nobile in the aftermath of the *Italia*'s wreck. Amundsen and a French crew of five disappeared on June 18, 1928. A wing float and fuel tank from Amundsen's Latham 47 "flying boat" were later recovered off the coast of Norway's Tromsoe Island, but

---

[6] The Mariner Museum online library: http://ageofex.marinersmuseum.org/

subsequent searches (most recently by a remotely operated submersible in 2004) have failed to find remains of the plane and crew.

Soviet airman Alexandr Kuznetsov carried out the first undisputed walk to the North Pole in 1948. His team of Soviet researchers landed in three Lisunov Li-2, Soviet air force modified DC-3s. They touched down near the pole and walked the rest of the way. The team carried out soundings through the ice, revealing the existence of a submarine mountain ridge on the floor of the Arctic Sea. In 2007, Russia's *Arktika Expedition* made a crewed submersible descent to the ocean floor at the North Pole. The submersible was piloted by Anatoly Sagalevich, who—apart from holding the record for the deepest freshwater dive in Lake Baikal—also has a movie credit in James Cameron's film *Titanic*.

The excursions in the Northern Arctic, both successful and tragic, taught explorers how to navigate polar terrain. At the same time some explorers continued north, adventurers were also taking on the continent of Antarctica.

Austral Adventures

Ernest Shackleton carried out several of the first wide-ranging explorations of the Harsh Continent. Taking part in three British missions, Shackleton first attempted to find the South Pole with Robert Falcon Scott on Scott's *Discovery Expedition* of 1901-04. Edward Wilson accompanied Scott and Shackleton. The three set a record for southward exploration, reaching 82°17'S before turning back. The Discovery Expedition set up a permanent hut, now known as Discovery Hut, near the site of what is today McMurdo Base.

On his second Antarctic trip beginning in 1907, Shackleton's *British Antarctica Expedition* (1907-09, also known as the *Nimrod Expedition*) came within 180 kilometers of the pole and carried out extensive exploration of Ross Island, ascending Mount Erebus for the first time (see next chapter). Three of Shackleton's men, Edgeworth David, Douglas Mawson and Alistair Mackay, set out to find the South Magnetic Pole, one of the Holy Grails of Antarctic exploration. On January 15, Mawson calculated that the trio was within miles of the South Magnetic Pole. The team abandoned their heaviest provisions and equipment, and made a mad dash for the spot. On January 16, 1909, they erected the British flag, took a photo, and then hurried back for their supplies. Once safely at home, they began to have doubts about their discovery. Mawson realized that he had not taken into account some new computations done by other researchers. But they had come close. Although the explorers fell short of their goal, the expedition gathered valuable scientific data throughout.

Just two years later, the Japanese mounted their first expedition to survey the King Edward II Land and the Alexandra mountains, but the geographic South Pole still beckoned. In 1911, a British party led by Robert Falcon Scott raced to beat Norway's Roald Amundsen to the South Pole. It was dubbed the *Terra Nova Expedition*. Scott had been to Antarctica once already on Britain's Discovery Expedition (1901-1904). But with news that Shackleton's group had narrowly failed to reach the South Pole, Scott was ready to try again.

Fig. 1.4. The ice-bound Endurance shortly before the sea ice punctured its hull, sending the ship to the floor of the Weddell Sea. Photo by Frank Hurley (Wikipedia Commons)

The Terra Nova Expedition, (more officially known as the *British Antarctic Expedition*) was backed by the Royal Geographic Society and private funding. Scott's party set out across the Harsh Continent with 65 men. Chief scientist Edward Wilson was also an accomplished artist, continuing a rich historical tradition of artists on the front lines of exploration. Scott's group carried out an extensive scientific program that included zoology, meteorology, physics and geology. While Scott's men settled on McMurdo Sound in their prefabricated hut waiting for the winter to end, they got word that Amundsen had arrived down the coast, planning an assault on the pole as well.

Amundsen, who had spent so much time in the arctic, traveled with only four compatriots and reached the South Pole by sledge and dogsled on December 14, 1911. He took an untried route that required only 57 days of travel. Along with colleagues Bjaaland, Wisting, Hassel and Hanssen, the Norwegian used dogsleds, skis and human sledges to reach his destination by that date. While the voyage was at times grueling, all members survived the return trip home.

Robert Scott's party followed a different route, covering more territory. His team consisted of himself and four others: Edward Wilson, Edgar Evans, Henry Bowers and Lawrence Oates. Using ponies for the initial part of the journey, the quartet at last reached the pole several weeks later, only to see the Norwegian flag flying before them. Amundsen had even left letters for the British team, along with one addressed to King Haakon of Norway. Scott's group perished on the return journey. Nevertheless, Robert Falcon Scott stands alongside Amundsen as a giant of Antarctic exploration. A memorial to those lost in the Terra Nova Expedition was erected in 1913 on Observation Hill, which today overlooks McMurdo Base. The cross serves as a reminder of the treacherous conditions on the Harsh Continent.

There was, of course, much more to Antarctica than the South Pole. While Scott and Amundsen prepared – with great fanfare – to get to the South Pole, Welsh-Australian geologist Douglas Mawson had other, more quiet plans. Mawson had been a member of the first party to climb Erebus in 1908 and had turned down the chance to attempt to reach the pole with Scott, instead choosing to lead the *Australasia Antarctica Expedition*. Mawson's goal was to chart 3,000 kilometers of Antarctic coast that lay south of Australia. He had the backing of Australia's Association for the Advancement of Sciences, as well as financial support from Britain's Royal Geographic Society and a host of private citizens. Mawson's 1911-1912 voyage of discovery netted much science, but was a harrowing endurance test of men and equipment. The expedition strategy saw Mawson's crew split into multiple groups with a total of eight separate forays. One group tended the base while the other three ventured into Antarctica's deep interior for their scientific research. Mawson took the lead of the party destined to survey several remote glaciers. The dangerous trek would be the farthest to travel and would require the heaviest loads, all carried on dog-pulled sledges. This team called itself the Far Eastern Shore Party.

Mawson chose British Lieutenant Belgrave Ninnis and Swiss lawyer and cross-country skier Xavier Mertz as his companions. 16 huskies pulled three sledges with a total of 1,720 pounds of gear and science equipment. In the course of travel, Ninnis broke through the ice into a crevasse. Two dogs died and Ninnis vanished completely, as did the majority of the expedition's food, along with the tent. The voyage that followed was a nightmare of starvation and bitter cold as the two survivors made the best of what was left and turned back for home. Mertz and Mawson both became ill, and Mertz died after a haze of fevered hallucinations in which he broke one of two remaining tent poles. Mawson, alone and miserable and having survived two crevasse falls, eventually made it back to base camp. He arrived on February 8, 1912, just in time to see the ship *Aurora* departing for Australia. Fortunately, a shore party had stayed behind to await the arrival of the Far Eastern Shore Party, but the ship was too far to call back. Mawson, sole survivor of his expedition,

Fig. 1.5. Grotto in an Iceberg; photo taken by Herbert Ponting during Scott's 1911 *Terra Nova Expedition* (Wikipedia Commons)

remained through another long Antarctic winter. What saved him in the end was the precision of his superb navigation and the patience of the shore party who stayed with him, nursing him back to health for the return trip to Australia the following summer. Mawson is considered one of the champions of the Heroic Age of Exploration.

Shackleton and the End of an Era

Ernest Shackleton followed Nimrod with the *Imperial Trans-Antarctic Expedition* from 1914 to 1917. Shackleton's vision was daring: send two parties to opposite coasts of Antarctica. One team, the Ross Sea party aboard the ship *Aurora*, would land at McMurdo Sound, where they would stage forays to lay supply depots across the ice and up the Beardmore glacier. Meanwhile, Shackleton's team would take the ship *Endurance* to make landfall upon the opposite side of the continent at Vashel Bay. Shackleton's men would then trek across the entire continent to the South Pole and onward to the Ross Sea using the depots that had been left by the Ross sea party, as it would have been near-impossible to carry so many supplies across the continent.

Fig. 1.6. After successful journeys with both Scott's and Shackleton's Ross Sea party, the *Aurora* was torn from its anchor in a vicious Antarctic storm. Today, its anchor remains near the shore (Photo courtesy USAP library/Michael Lucibella)

The Ross sea party successfully deployed the supply caches, but a storm ripped the *Aurora* from its moorings. Ocean currents pulled the Aurora away from the sledding parties marooned on shore and drifted for over six months before breaking free of the ice, stranding many of the crew. The ship had a damaged rudder and was forced to return to New Zealand rather than return to pick up its shore party. All but three of the stranded crew were eventually rescued, but for a time they were forced to endure extreme weather and illness. Still, the group was able to carry out its mission laying the depots for Shackleton's party.

On the other side of the continent, *Endurance* became icebound. The shifting floes crushed the ship, and it sank. Shackleton's 28 members survived on the ice in improvised camps for several months, finally taking to the sea in three lifeboats. They arrived at the desolate Elephant Island in the South Shetlands, which afforded little more shelter than they had before. Realizing that the desolate island was far from shipping routes and that their chances of being rescued were slim, Shackleton and five other brave men set out in one of the three lifeboats, the *James Caird*, a 6.9 m (22.5 foot) open boat. Among the men were Irishman Tom Crean, who had been a member of the Discovery Expedition and Scott's Terra Nova Expedition. Shackleton was confident that the hardy Irishman could endure the voyage. Shacklenton also took accomplished navigator Frank Worsley, who would later write, "We knew it would be the hardest thing we had ever undertaken, for the Antarctic winter had set in, and we were about to cross one of the worst seas in the world." They braved 800 miles of open sea before reaching South Georgia Island, where they assembled a search party and rescued every member of the rest of Shackleton's crew. Shackleton and his team should have returned to a hero's welcome, but World War I had broken upon Europe, and their arrival was barely noted. History has looked kindly upon them: Shackleton's expedition is equated with bravery, ingenuity in the face of the harshest of conditions, and, yes, endurance. The *James Caird* lifeboat was returned to England and put on display at the National Maritime Museum in London, where the author Rosaly Lopes used to work. Looking at it, one is flabbergasted wondering just how Shackleton and his crew were able to make their journey in rough open seas. Nowadays, the little boat is exhibited at Dulwich College in London, Shackleton's old school, on a bed of stones that includes many from South Georgia Island.

Shackleton's and Mawson's expeditions may have marked an end to the "heroic" days of exploration, but as is often the case, they paved the way for future forays. It was not until 1928 that mechanized travel came to the Harsh Continent under US Rear Admiral Richard E. Byrd. He used airplanes, aerial photo reconnaissance, Ford-built snowmobiles and a network of communications devices to an extent unmatched by his predecessors. Byrd's group was the first in Antarctica to carry out regular wireless communications with the outside world.

Byrd mounted a second expedition incorporating the first seismic measurements of the Ross Ice Shelf, proving that the shelf was a floating ice field on the surface of the ocean. The expedition transmitted the first human voices, powered by the first motor-driven electrical generators. Byrd's group examined a variety of biology ranging from plankton to seals and documented changes in the Ross Ice Shelf since Scott's Terra Nova Expedition. Douglas Mawson's earlier (1912) expedition led to even further projects and resulted in a much larger foray sponsored by the governments of Australia, Britain and New Zealand. The project, known as BANZARE (for British, Australian and New Zealand Antarctic Research Expedition) involved two separate trips in 1929 and 1930. Many expeditions followed representing many different nations, but the prospect of voyaging through the southern continent remained a dangerous and audacious one. Those who dared to go needed more permanent bases of operation, and those were to come soon.

Highlights of Major Early Antarctic Explorations*

Date	Explorer/s	Comments
1675	Antonio de la Roche	Blown off course, he catches sight of South Georgia
1722	Yves Joseph de Kerguélen-Trémarec	Discovers Isles Kerguélen
1739	Jean-Baptiste Bouvet de Lozier	Discovers Bouvet Island
1773-1775	James Ross Cook	Crosses Antarctic circle, discovers South Sandwich Islands
1819	William Smith	Sights South Shetland Islands
1820	Edward Bransfield & William Smith	Scout south of Shetlands, first sight of Antarctic Peninsula?
1820	Fabian Gottlieb von Bellingshausen	First sighting of Antarctic continent
1821	John Davis	First landing on the Antarctic mainland
1821	Nathaniel Palmer/George Powell	Discovery of South Orkney Islands
1823	James Wedell	Sails to 74°S
1822	Benjamin Morell	First landfall on Bouvet Island
1831	John Biscoe	First Antarctic sighting from Indian Ocean side
1840	Charles Wilkes	Discovers Wilkes Land
1840	Jules-Sebastian Dumont d'Urville	Discovers/surveys Adélie Coast (named for his wife)
1841	James Clark Ross	Discovers Victoria Land, Ross Island, and Mt. Erebus
1892	Carl Larsen	First discovery of fossils indicating warmer earlier climate
1895	Henryk Bull, Carsten Borchgrevink	Find first indigenous plant life (lichens)
1898	Adrien de Gerlache	Stranded crew first to survive Antarctic winter
1902	Otto Nordenskjöld	First major sledge exploration (400 miles)
1902	Eric von Drygalski	Discovery of Wilhelm II Land/extensive science
1902	Robert Falcon Scott	3100 mile journey from McMurdo reaches 82°S
1904	Jean Baptiste Charcot	Extensive survey of western Antarctic Peninsula
1904	William S. Bruce	Discovery of Coats Land
1904	Carl Larsen	First whaling station established
1908	Shackleton, Wold, Marshall and Adams	South Pole attempt; team comes within 97 miles
1909	Douglas Mawson	Reaches South Magnetic Pole
1911	Nobu Shirase	First Japanese Antarctic Expedition
1911	Roald Amundsen	Expedition reaches South Pole
1912	Robert Falcon Scott	Reaches South Pole a month after Amundsen, all perish
1912	Wilhelm Filchner	Discovers Luitpold Coast
1912	Douglas Mawson	First radio in Antarctica, discovery of new coastal area
1915	Ernest Shackleton	Two year ordeal ends in heroic survival
1928	Hubert Wilkins	First flight in the region
1929	Richard E. Byrd	First to fly over South Pole
1947	Operation Highjump	4700 US Navy personnel set up permanent base
1950	Sweden, Great Britain, Norway	International expedition in Dronning Maud Land
1957	International Geophysical Year	67 countries establish bases across the continent

*Note: Up to 1959, over 300 expeditions were carried out on the mainland and in the vicinity of Antarctica. This chart covers only a portion of them.

Permanent Beachheads

Gaining a foothold in a place as desolate as Antarctica is a formidable challenge. The first attempt was made by Carsten Borchgrevink and his crew of the *Southern Cross*, a Norwegian whaling vessel. The adventurers set up Camp Ridley (named after Borchgrevink's mother) on the shore of Cape Adare in February of 1899. The beachhead presented a bleak site, according to the diary of physicist/astronomer Louis Bernacchi: "Approaching this sinister coast for the first time…our decks covered with drift snow and frozen sea water, the rigging encased in ice, the heavens as black as death…struck into our hearts a feeling preciously akin to fear … It was a scene, terrible in its austerity, that can only be witnessed at that extremity of the globe; truly, a land of unsurpassed desolation.[7]" The ten men erected prefabricated huts and brought 75 sledge dogs to the forbidding coast. They were the first to spend a winter on the mainland. The camp was hardly an establishment of civilization, nor was it designed to be.

The hazardous environment of Antarctica demands that if there is to be any reliability in the survival of its explorers, there must be some infrastructure set down. The earliest substantial bases established on the continent were military. During World War II, the British launched Operation Tabarin in response to German and Argentine territorial claims to the southern continent and its outlying islands. British officer James Marr, who had been a member of Shackleton's Trans-Antarctic Expedition, led a party of 14 that departed the Falkland Islands in 1944. They established the first permanent base near an abandoned Norwegian whaling station on Deception Island. Two other bases were set down at Port Lockroy and Hope Bay, the first on the Antarctic mainland. In 1948, Chile established its first outpost. This was quickly followed by the United Kingdom's Signy Research Station, Australia's Mawson Station, and the French Dumont d'Urville Station, all in 1956. With the Cold War heating up, the two superpowers established posts the same year. The Soviets founded their Mirny Station on Queen Mary Land, and the US built McMurdo Station on the south end of Ross Island. These science stations were followed the next year by several others, including New Zealand's Scott Base (named for Robert Falcon Scott). Since then, some 70 stations and remote outposts have sprung up across the desolate Antarctic wilderness. Even tropical Brazil hosts a science station, the northernmost on the continent.

The largest base in Antarctica, McMurdo Station, gets its name from British naval vice-admiral Archibald McMurdo, whose maritime career included two expeditions aboard the HMS *Terror*, a British warship converted for scientific exploration. The first voyage surveyed the coastline and environs north of Hudson Bay. On McMurdo's second voyage, the newly advanced second-lieutenant commanded the *Terror* and accompanied James Clark Ross aboard the sister ship *Erebus* to Antarctica in 1839-1843. His name has also christened McMurdo Sound (where McMurdo Station sits), McMurdo Ice Shelf, the McMurdo Dry Valleys and the McMurdo-South Pole Highway, which links McMurdo Station to the South Pole Station some 1,600 kilometers away. Ironically, while expedition leader Ross' name is associated with the island and ice shelf, McMurdo's name became far more famous, as McMurdo Sound became the staging site for South Polar Expeditions and other forays across the Antarctic wilderness.

---

[7] See *Physics in Australia to 1945* - BERNACCHI, Louis Charles at www.asap.unimelb.edu.au

International Geophysical Year

Many of Antarctica's facilities were established during the 1957-1958 International Geophysical Year (IGY). 12 nations instigated the IGY, which included 67 Antarctic stations. Officially beginning on July 1, 1957, the "year" closed December 31, 1958 and marked an unprecedented cooperation among nations that had been reticent to exchange information at the height of the Cold War (particularly between Eastern and Western countries). Promoting exploration and research, the IGY instigated a push to explore and set up outposts in the Antarctic interior, an area that until then had scarcely been explored.

The IGY enabled the establishment of the first year-round research stations in the interior of the continent. According to the National Science Foundation, the IGY was the "greatest coordinated scientific assault on Antarctica ever mounted." The United States carried out a full court press, setting up seven winter-safe stations. Researchers founded four in coastal regions (Little America, Hallett, Wilkes Station and Ellsworth). Two bases were established deep in the continent's interior. The Byrd Station consisted of four prefabricated buildings, later replaced by an underground post in 1960. The year-round station was closed in 1972 and became a summer field camp. A sophisticated base was also erected at the South Pole. The Amundsen-Scott South Pole Station sits atop the Antarctic Ice at 2,835 meters (9,300 feet) above sea level. The original structure (now affectionately referred to as "Old Pole") consisted of a group of prefabricated buildings assembled by a US Navy crew. By 1960, most of the buildings had been swamped by snow, and the site had to be completely abandoned in 1975. At that point, engineers constructed a geodesic dome, which has since been replaced by a more modern set of structures.

Major IGY facilities were erected by Britain's Royal Society (Halley Research Station), France (Dumont d'Urville Station), Belgium (King Baudoin Base), and Japan (Showa Station) among others. Japan also contributed its icebreaker Soya to be a South Pole observing ship.

New Zealand and the US set up a joint scientific post at Cape Hallett. The station was in year-round use until 1964, when fire damaged it (limited use continued until 1973). As McMurdo was set up as a logistics center for the IGY but not as a science center, New Zealand took the lead as the network linking Hallett and other science facilities. At the conclusion of the IGY, New Zealand decided to continue research at Scott Base and continues to do so today in varied fields including aurorae, geophysics, ionospheric and magnetospheric research, as well as a wide array of marine, biological and geological studies. The peak summer population at New Zealand's Scott Base is about 200, dropping to below 50 in the long, dark winter.

The largest construction project of the IGY's Antarctic arena was the establishment of the US McMurdo Station (see Chapter 4). McMurdo, perched on McMurdo Sound, was set up in anticipation of the IGY as a logistics and supply base for more distant field camps and science bases. Construction actually began in 1955 adjacent to the Discovery Hut built by Robert Falcon Scott in 1902. Today, McMurdo stands as the largest settlement in Antarctica, housing over 1,000 people at the height of the austral summer season. Its campus consists of about 85 buildings, ranging in size from three-story edifices to small radio shacks and domed observatories. Garages repair and maintain the many vehicles that support McMurdo and surrounding areas. College-style dormitories, administrative buildings, an extensive galley, several coffee houses/pubs, a firehouse, and a general store all provide welcome infrastructure to visiting scientists and support staff. McMurdo's power plant and a seawater distillation facility service the famous Crary Science Laboratory, where researchers are supplied with laboratories and equipment to work locally. Above-ground water, sewer, telephone, and power lines link the entire McMurdo settlement. McMurdo continues to support year-round research, serving as the primary logistics facility for resupply of inland stations and remote field camps.

Fig. 1.7. The small, prefabricated structures of McMurdo Station as they stood in 1956 when the facility was known as Naval Air Facility McMurdo (USAP photo by Dave Grisez), and in 2017 (Photo by the authors)

McMurdo Station was originally called Naval Air Facility McMurdo. Seabees erected many of its structures, including a nuclear reactor that went on line in 1962. The nuclear plant was shut down in 1972 in favor of diesel-powered generators. Wind turbines and local solar panels round out McMurdo's present power needs.

The construction carried out during the International Geophysical Year was a means to an end, and that end was science. Data flowing from the international cadre of facilities and scientists gave the world its first in-depth overview of the workings of the Antarctic continent. Areas of study included aurora and airglow, magnetospheric investigations, cosmic rays and astronomy, meteorology, oceanography, seismology and solar studies.

International Antarctic Treaty

The 12 nations that took part in the International Geophysical Year participated in the world's first treaty pertaining to Antarctica. The landmark declaration sets all territory south of the 60°S latitude as a peaceful zone of international cooperation. Drafted at the height of the Cold War, the remarkable treaty prohibits military activity, including construction of military installations of any kind. At the same time, the treaty allows for the use of military personnel and equipment in the aid of research or other peaceful purposes.

Fig. 1.8. Painting done on-site of McMurdo's *Chapel of the Snows* (Acrylic on canvas by Michael Carroll)

The International Antarctic Treaty also bans nuclear explosions or disposal of radioactive waste. By international agreement, mining is also prohibited. All signatories of the treaty are to be given full access to inspect any site in order to ensure adherence to all aspects of the treaty by all partners. At the same time, the document is designed to foster scientific research, exchange of results, and international cooperation on other fronts. In fact, the modern history of Antarctica serves as a shining example of international cooperation and the sharing of resources, talent and infrastructure. An example is the work done between New Zealand's Scott Base and the US McMurdo Station, where wind turbines contribute 30% of electricity to both facilities. The two centers also share a helicopter landing base and the William's and Phoenix airfields.

As of 2007, 16 additional nations had joined with "consultative" status, meaning that they accede to the Treaty's specifications and actively carry out extensive science. Another 22 nations have signed on as observers, agreeing to the Treaty and attending meetings as non-voting members. Today, some 50 countries embrace the treaty, representing two thirds of the world's population.

According to the treaty's provisions, no single nation can claim real estate on the southern continent. When one arrives in Antarctica, no passport is required. Marriages cannot be performed there, as the territory is not associated with any specific country. Marriage ceremonies have been performed at McMurdo in its chapel, the *Chapel of the Snows*, but these are informal and not internationally recognized. According to the treaty, "No acts or activities…shall constitute a basis for asserting, supporting or denying a claim to territorial sovereignty in Antarctica. No new claim, or enlargement of an existing claim, to territorial sovereignty shall be asserted while the present Treaty is in force."

Along with Antarctica, three regions are identified as "global commons" by treaties: the high seas, the Earth's atmosphere and outer space. These were defined as spaces outside the territory of nations and beyond normal inhabitable zones for the human species. The Outer Space Treaty (1967), which forms the basis of Space Law, derives its origin from the International Geophysical Year, as does the United Nations Convention on the Law of the Sea (1958) and the Antarctic Treaty (1959). The treaties all share legal principles for the exploration and governance of these "no-man's lands," with the aim of safeguarding and protecting the specified regions whilst avoiding political tensions. Antarctica provided a model for the treaties, and was a test case for how international collaboration on extraterrestrial bodies might be done.

Antarctic Conservation Act

Beginning in 1964, a series of international agreements formalized measures to protect the delicate Antarctic environment. While its frozen landscapes may appear sterile, life does exist even on the land, and where it hangs on, it is fragile. In places where microbes are rare, there is no biological breakdown of waste or toxins, so these must be handled carefully. The richest life is along the coasts and in the water, where penguins, whales and seals live side by side with a wide array of fish, crustaceans and other sea life. In the 1960s, ships began taking tourists into Antarctic waters where they could interact with the wildlife, usually in uncontrolled ways. Overflights from Australia and New Zealand commenced in the early 1970s. With these trends in mind, the 1964 *Agreed Measures for the Conservation of Antarctic Fauna and Flora* passed among the signatories of the International Antarctic Treaty. In 1991, this was expanded by the *1991 Protocol on Environmental Protection* and the *Antarctic Science, Tourism and Conservation Act of 1996*. Article 2 of the 1991 Protocol classifies Antarctica as a "natural reserve, devoted to peace and science."

All nations that take part in Antarctic exploration adhere to these treaties and have crafted their own guidelines for their respective citizens. In 1978, the United States passed its *Antarctic Conservation Act (ACA)*, which applies to not only all US citizens but also to any Antarctic expeditions – scientific or tourist – that originate from the United States. In conjunction with international treaties, the ACA seeks to minimize pollution and protect native wildlife and plants. It requires a specific scientific permit for otherwise forbidden activities. These include the capture of native mammals and birds, the engagement of any harmful interference with wildlife, the entrance to Antarctic Specially Protected Areas (ASPAs), and the introduction of species or waste materials into the Antarctic environment. Fines of up to $28,000 (USD) and one-year imprisonment for each infraction may be leveled at violators. Soil may not be introduced into Antarctica unless it is sterile. (Many Antarctic facilities raise food in hydroponic gardens.) Domestic edible plants or laboratory animals and plants (including microbes and fungi) may only be introduced under special permit, and permit applicants must be able to demonstrate a clear need.

All cruise lines are now required to have permits before any tourists may land on the mainland or surrounding islands. Most cruise ships stage out of Ushuaia, Argentina. Their landings usually take place by water (with the use of inflatable Zodiacs) or by helicopter. The majority of tourists visit Antarctica under the umbrella of the International Association of Antarctic Tour Operators, an organization dedicated to promoting safety and environmental accountability within the community of tour operators and guides.

Antarctica has served as a prime example in the study of many proposed international governing models, from inhabiting the ocean floor to settling territory among the planets and moons. As one McMurdo worker put it, "When you're in a place like this, sometimes in survival mode, you don't argue about who owns what rock."

The United States Antarctic Program (USAP)

The IGY jump-started continent-wide infrastructure across the Antarctic mainland and outlying islands. Exploration and travel across the continent has evolved from sporadic, dangerous journeys by foot and dogsled to more regular travel via land, sea and air. The National Science Foundation and other organizations have put in place multiple ways to make the Harsh Continent a safer dwelling. Through their efforts, hundreds of people are able to carry out research annually in the most remote corners of Antarctica's frozen wilderness, from hunting for meteorites in glacial fields to researching extremophiles in the Mars-like Dry Valleys. Although flights to and within Antarctica are often cancelled, as they are still at the mercy of Antarctica's capricious weather, transport and communication in the austral lands is fundamentally different from the time Shackleton took out his advertisement. Currently, a significant part of the exploration of Antarctica falls under the auspices and funding of the National Science Foundation's United States Antarctic Program (USAP), established in 1959. The majority of travel and logistics for USAP's supported personnel is staged out of McMurdo Station. Although it began as a naval logistics center, the base is now managed by this program. Operations were turned over to the National Science Foundation in the aftermath of the IGY, as science results from the year-long international study proved so promising that many member nations decided to continue Antarctic research. Under the patronage of USAP, scientific disciplines like biology and ocean sciences were added to the list of research areas already being pursued. NSF tasked the Defense Department with sustaining the extensive science logistics needed in the southern camps and outposts supported by McMurdo.

After 1971, the National Science Foundation assumed overall responsibility of all US activities on the Antarctic continent and its environs. The scope of research carried out today has expanded greatly with improved infrastructure of the area. Investigators from across the US pursue the USAP's goals of upholding the Antarctic Treaty, nurturing collaborations with scientists throughout the world, safeguarding the delicate Antarctic environment and developing strategies to ensure that Antarctica's limited assets are used wisely and in ways fair to the global community.

Roughly 3,000 American scientists and support staff take part in USAP operations each year. These operations overlap with the work of other nations, particularly in the area of transportation. The workhorse of the Antarctic skies, the ski-equipped Hercules LC-130 aircraft, provide transport of equipment, supplies and personnel across the continent. Larger C-17 transports fly early in the season when the ice is strongest, but during summer months the smaller Hercules fly. The US planes are crewed by the US Air National Guard. Helicopters and small aircraft flown by contractors like Lockheed Martin provide support for research teams in the field. De Havilland Twin Otters and modified DC-3s, called Basler BT-67s, are piloted by Canadian, British and New Zealand teams. Kenn Borek Air

Ltd. – based in Calgary – offers local cargo and passenger service to science teams in the field for research, medical runs (including evacuations) and resupply.

Three airfields service Antarctic operations for McMurdo. An **ice runway** on the McMurdo Sound sea ice near McMurdo Station is used sporadically by small, wheeled aircraft in October, November and early December.

Fig. 1.9. The buildings of Williams Airfield rest on skids for easy movement (Photo by the authors)

Later in the summer, temperatures rise and weaken the sea ice. Flight operations shift to **Williams Field**, a 3,000-meter snow/ice runway on the Ross Ice Shelf. Also known as Willy Field, it was named for Richard T. Williams, who died when his 30-ton tractor broke through the sea ice during Operation Deep Freeze, a US Navy project to support the building and supply of US facilities established for the IGY. The buildings and support structures at Williams Field are mounted on sleds for rapid deployment and relocation. There's even an airport shuttle; a large, wheeled "TerraBus" ferries passengers 16 kilometers to the nearby McMurdo Station. It is nicknamed "Ivan the TerraBus."

A third airfield, the **Pegasus ice runway** (named for *Pegasus*, a US Navy C-121 Super Constellation that crashed nearby without casualties during the 1970/71 season), was about 25 km from McMurdo. In recent years, higher temperatures and black dust from nearby Black Island (named such by the Discovery Expedition because of its lack of snow) damaged the runway, and it has now been abandoned in favor of a new facility three miles to the northeast, called Phoenix Airfield. Rather than relying on existing ice, the Phoenix site consists of compacted snow utilizing advanced construction techniques. The new runway, out of the wind track from Black Island, is projected to handle 60 wheeled flights every year, along with other ski-equipped landings. The new site opened in the 2016/17 season.

Fig. 1.10. Workhorses of the Antarctic skies (*L* to *R*): the Bassler BT-67, the Twin Otter, and the Bell 212 helicopter (Photos by the authors)

Antarctic Travel in the Modern Age

We've come a long way since the early Antarctic maps with large swaths labeled "unknown regions." Just as people from Michigan hold up their flattened right hand as a map of their state (with Detroit at the base of the thumb and the fingers pointing northward), so Antarctic veterans have a "handy" map wherever they go (Figure 1.11). In this case, the left hand with fingers folded into a fist and the thumb outstretched (like a hitchhiker) serves as a roadmap for polar landmarks. The tops of the folded fingers represent the polar plateau, which includes the South Pole at about the tip of the middle finger. The Transantarctic Mountain Range follows the tips fingertips across the upper palm. The lower left of the palm outlines West Antarctica, with the Ross Sea Ice Shelf below, putting McMurdo near the end of the pinkie finger. The thumb, representing the Antarctic Peninsula, points toward South America, while the wrist heads off toward New Zealand. Even seasoned McMurdo inhabitants can be seen holding their fist up to communicate the latest news at specific locations on their portable maps.

Most researchers and staff bound for Antarctica travel by air. 30 international outposts and bases operate landing sites for either helicopters or aircraft. 27 bases have helicopter pads, and another 15 have runways of gravel, ice or compacted snow.

Land travel is necessary but more hazardous. McMurdo's tracked and wheeled vehicles deliver supplies across treacherous landscapes, over mountains, and through ice and snow. There are few roads in Antarctica, and those that exist near major stations are rugged with severe polar weather preventing upkeep beyond the summer season. Routes through snowy terrain are marked by long parades of flags. Red and green flags, each about ten meters apart, serve as a way back home if a whiteout occurs. Blue flags mark fuel lines, which snake across the surface of the ice. Black flags indicate dangers such as thin sea ice or crevasses.

Vehicles come in a variety of shapes and sizes, manufactured in many different countries. Fairly conventional 4WD vehicles are sufficient for some areas. Others require tracked vehicles like the Swedish Haaglunds, a two-trailer amphibious vehicle designed to float should it break through the sea ice.

Perhaps the most dramatic regularly scheduled land voyage travels along the route known as the South Pole Overland Traverse route, affectionately called "SPoT." Stretching across 1,450 km (900 miles) of frozen wilderness, the path links McMurdo Station to the

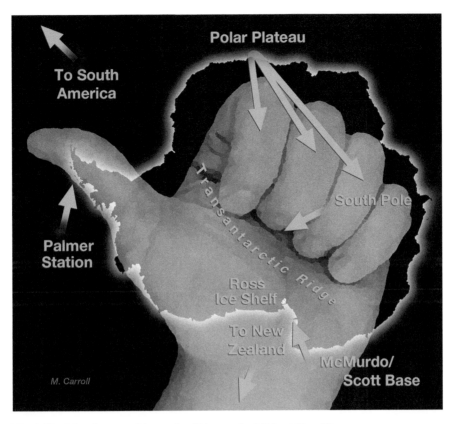

Fig. 1.11. A handy map of Antarctica (Diagram by Michael Carroll)

Amundsen-Scott South Pole Station. In the past, fuel was flown to the pole aboard Hercules LC-130s, but the unpredictable weather often precluded full recharging of Antarctica's hungry infrastructure. Instead, SPoT's massive tractor-trailers, riding on overgrown tank treads, now transport bladders containing 100,000 gallons of fuel. LC 130s require dozens of flights to deliver the same amount of fuel, use more fuel than the Caterpillar tractors involved in the SPoTs, and leave a much larger carbon footprint.

As with almost every operation at the pole, the SPoTs must first go through McMurdo.

Each year, an advanced party must plow the way, locating and filling in crevasses and mounting the Leverett Glacier of the Transantarctic Mountain range, then on up to the Polar Plateau. The grueling journey is followed by one or often two successive caravans ferrying supplies of fuel and food to the remote base at the pole. In the 2016/17 season, SPoT1, SPoT2, and SPoT3 delivered a combined total of 310,000 gallons of fuel, equivalent to 60 supply flights of Hercules aircraft. In addition to supply carriers, the tractors pull bunkhouse sleds, a kitchen, and a garbage/waste sled. The roughly 40-day trip to the pole is far longer than the return trip, which carries less weight.

Fig. 1.12. The amphibious Haaglunds vehicles serve a variety of functions in the Antarctic environment (Photo by Rosaly Lopes)

Fig. 1.13. Fuel is carried in giant bladders during the SPoT transports (USAP photo by Tim Thomas)

McMurdo serves as Antarctica's major seaport. Although many coastal outposts have docks or offshore anchorages, McMurdo services large cargo vessels. These ships deliver supplies and carry out waste to be disposed of at facilities in California. Small boats and barges ferry passengers and supplies from ships to shore, as do helicopters. Major shipping to the continent typically stages through Punta Arenas, Argentina.

The US Antarctica Program (USAP): Raison d'etre

Research carried out by the USAP has three goals. The first is to better understand the continent in terms of its ecology and environment. The second is to better comprehend Antarctica's effects on and responses to global processes like long-term meteorology, ocean temperatures and currents, and global climate patterns. The third is to use the southern continent as a base for research into the upper atmosphere and space environments. Antarctica is uniquely suited for many of these studies, but its isolation and climatic conditions make field research costly. Research is carried out in Antarctica only when it cannot be performed at less hostile sites.

The USAP hosts three year-round bases at McMurdo, the South Pole Station, and Palmer Station. Remote camps scatter across the continent in such remote locations as the Dry Valleys, along coastlines, and upon mountaintops including Erebus.

These widely varied centers carry out a broad spectrum of science from astronomy to paleontology. But the rewards to the global village go far beyond theoretical studies of obscure scientific projects. The science taking place on the southern continent has practical value to everyday life and may in fact provide insights that have world-changing effects. Lessons and technology development from the Harsh Continent enrich fields as diverse as biology to world environment to medicine. Antarctica's unique platform enables scientists to carry out research not possible on other parts of the globe.

Most civilians will never have the opportunity to see the sweeping snowscapes, surreal ice towers, and vast sheets of Antarctic's sea ice and glaciers. Fewer still have a sense of the value of science in the everyday life and, ultimately, to the well-being of humankind at large. Through literary, visual, and other arts, the USAP Artists and Writers program seeks to teach taxpayers how their monetary contributions are being turned into research and new scientific knowledge. Since the main purpose of the U.S. Antarctic Program is research and education, the Antarctic Artists and Writers Program sponsors writing and artistic projects that are designed, according to the National Science Foundation, "to increase the public's understanding and appreciation of the Antarctic and human endeavors on the southernmost continent." The NSF gives priority to projects that are specifically engineered to interpret and exemplify the scientific pursuits being conducted in or around the Antarctic region. According to NSF's Dr. Peter West, the agency sends just three or four artists to Antarctica each summer season to "help communicate the significance of the science done there—and the challenges of working in that hostile environment to do science successfully—to the general public. Many scientists are good communicators, but art reaches people in a different way than data or graphs can."

Participants in the program are chosen through exactly the same process that scientists are, a competitive review of the proposal by an external panel of experts. The major difference is that Artists and Writers are not given a grant as scientists are, except for a token $1 to the Principal Investigator, in our case Michael Carroll. Rather, NSF housed and fed us, supplied us with Big Reds and other formidable sets of standard Extreme Weather Clothing, provided us with the assistance of an expert mountaineer, and got us into the field to do our work.

Artists in exploration

Art and exploration have gone hand in hand throughout the modern era of exploration. When John Lloyd Stephens (1805-1852) set out on his series of groundbreaking expeditions to the Mesoamerican ruins of Mexico and Central America, he insisted on bringing along artist/architect Frederick Catherwood. In Stevens' words, Catherwood's work could "provide the world a visual sense of the wonders we had seen…" Artists frequently accompanied early European expeditions into the American West. The paintings and sketches of Thomas Moran and Albert Bierstadt helped to convince the US Congress to establish the first National Parks at Yosemite and Yellowstone.

Artists have been an integral part of Antarctic exploration since the days of Shackleton and Scott.

With the advent of videography and photography, explorers bring new art forms to bear witness to today's explorations. And while photography provides visual inspiration as well as accurate documentation, the fine art of painting can complement this form with its nuanced focus and emotional content. The paintings of our expedition to Erebus will continue the great tradition of artists like Moran, Bierstadt, Alaska's Sir Sydney Lawrence, explorer Frederick Church and others.

Fig. 1.14. (*Left*) *H.M.S. Challenger* artist John James Wild painting on Antarctica's Kerguelen Island (Courtesy ©Natural History Museum, London/Science Photo Library); (*Right*) Michael Carroll painting in considerably more comfortable conditions in the garage of the Lower Erebus Hut (Photo courtesy Nial Peters)

Why Go Today?

One must have a very good reason for going to Antarctica, and an even better reason for going to the Mt. Erebus volcano on Ross Island as guests of the National Science Foundation's *Artists and Writers program*. Specifically, we journeyed to Antarctica in search of planetary analogs. The concept is one of the most important ones in the study of distant worlds: the Earth exhibits terrain and geology that bears strong resemblance to features we see on other planets and moons. For example, the layered terrain of the Martian poles, put down by periodic dust storms and seasonal winds, looks a lot like the layered terrains in the Icelandic glacier Vatnajokull, where periodic volcanic eruptions drape the ice with ash before winter ices lock them into place. Many such analogs exist in Antarctica. Its dry valleys are famous for their Mars-like conditions. Sheltered from the Antarctic Ice Sheet by the Transantarctic Mountains, these arid, chilled hollows host subsurface ice, frozen lakes, and rocky gravel, making them the closest Martian analog on the planet and a prime target for astrobiologists searching for hints about life in the cosmos. Antarctica also offers analogs to locales farther afield: fractures and pressure ridges sculpt the sea ice in ways reminiscent of Jupiter's oceanic moon Europa. Mount Erebus, our prime target, builds eerie ice towers around vents on its flanks, perhaps presaging volcanic forms that future explorers will find by the geysers of Enceladus or Triton. In addition, Erebus is one of only half a dozen volcanoes on Earth with currently active lava lakes, making it a close cousin to Jupiter's violent, volcanic moon Io. And while these analogs are valuable to planetary scientists, they also serve as a critical tool for astronomical artists, who depict alien vistas at resolutions not yet available from spacecraft.

Fig. 1.15. (*Left*) Iceland's Vatnajokull glacier contains layers of volcanic dust interlaced with ice (USGS); (*Center*) Ground view of ice layered with sulfur and ash in the Hekla region (Photo by Marilyn Flynn); (*Right*) Periodic seasonal dust storms on Mars leave similar layered terrain as they are locked into the polar ices each winter (Mars Reconnaissance Orbiter image courtesy NASA/JPL/University of Arizona)

The study of planetary analogs is an urgent matter. The lessons we uncover help us to better understand our own world and serve as cautionary tales for our care of Earth's biome. Studies of Jupiter's colorful cloud bands have given meteorologists insights into the working of large weather systems like hurricanes. Studies of the Martian polar caps yield portraits of climate change on a world that was once much more like the Earth than it is today. The Venusian atmosphere has provided us with critical information about ozone-damaging chemicals like chlorofluorocarbons. In many important ways, our study of planets and moons has been a study of the Earth.

With recent discoveries of volcanoes and geysers on other planets and moons, Antarctica's Erebus undoubtedly provides insights into the exploding mountains and spewing geysers across the Solar System. But how closely does Erebus resemble a Martian volcano? How similar are its ice towers to structures we may see on the moons of the outer planets? How much of Io simmers in its lava lake? To answer these questions, we had to understand the nature of Mount Erebus itself.

Fig. 1.16. South Korea's Jang Bogo Station is the third largest permanent facility in Antarctica (Photo courtesy NASA/Christine Dow)

Antarctic Science Stations

Base/outpost	Hosted by	Established	Location
Aboa*	Finland	1989	Queen Maud Land
Almirante Brown Antarctic Base*	Argentina	1951	Antarctic Peninsula
Amundsen-Scott South Pole Station	United States	1957	Geographic South Pole

Base/outpost	Hosted by	Established	Location
Artigas Base	Uruguay	1984	King George Island
Asuka Station*	Japan	1985	Queen Maud Land
Belgrano II	Argentina	1979	Coats Land
Bellinghausen Station	Russia	1968	King George Island
Bernardo O'Higgins Station	Chile/Germany	1948 1991	Antarctic Peninsula
Bharati	India	2012	Larsemann Hills
Byrd Station*	United States	1957	Marie Byrd Land
Camara Base*	Argentina	1953	Half Moon Island
Captain Arturo Prat Base	Chile	1947	Greenwich Island
Carlini Base	Argentina	1953	King George Island
Casey Station	Australia	1957	Vincennes Bay
Comandante Ferraz Antarctic Station	Brazil	1984	King George Island
Concordia Station	Italy/France	2005	Dome C, Antarctic Plateau
Dakshin Gangotri	India	1983	Dakshin Gangotri Glacier
Davis Station	Australia	1957	Princess Elizabeth Island
Deception Station*	Argentina	1948	Deception Island
Dome Fuji Station*	Japan	1995	Queen Maud Land
Drushnaya 4*	Russia	1987	Princess Elizabeth Land
Dumont d'Urville Station	France	1956	Adelie Land
Base Presidente Eduardo Frei Montalva and Villa Las Estrellas	Chile	1969	King George Island
Esperanza Base	Argentina	1953	Hope Bay
Gabriel de Castilla Base*	Spain	1989	Deception Island
Gonzalez Videla Antarctic Base*	Chile	1951	Paradise Bay
Great Wall Station	China	1985	King George Island
Halley Research Station	UK	1956	Brunt Ice Shelf
Henryk Arctowski Polish Antarctic Station	Poland	1977	King George Island
Jang Bogo Station	South Korea	2014	Terra Nova Bay
Jinnah Antarctic Station*	Pakistan	1991	Sor Rondane Mountains (Queen Maud Land)
Juan Carlos Station*	Spain	1988	Livingston Island
King Sejong Station	South Korea	1988	King George Island
Kohnen Station*	Germany	2001	Queen Maud Land
Kunlun Station*	China	2009	Dome A (East Antarctica)
Law-Racovita Station	Romania	1986	Larsemann Hills (Princess Elizabeth Land)
Leningradskaya Station*	Russia	1971	Oates Coast, Victoria Land
Machu Picchu Base*	Peru	1989	King George Island
Maitri Station	India	1989	Schirmacher Oasis
Maldonado Base*	Ecuador	1990	Greenwich Island

Base/outpost	Hosted by	Established	Location
Marambio Base	Argentina	1969	Seymour-Marambio Island
Mario Zucchelli Station	Italy	1986	Terra Nova Bay (Ross Sea)
Matienzo Base*	Argentina	1961	Graham Land
Mawson Station	Australia	1954	Mac. Robertson Land
McMurdo Station	US	1956	Ross Island
Melchior Base*	Argentina	1947	Melchior Islands
Mendel Polar Station*	Czech Republic	2006	Ross Island
Mirny Station	Russia	1956	Davis Sea
Mizuho Station*	Japan	1970-1987	Mizuho Plateau
Molodyozhnaya* Station	Russia/Belarus	1962	Thala Hills
Neumayer Station III	Germany	2009	Atka Bay
Novolazarevskaya Station	Russia	1961	Queen Maud Land
Orcadas Base	Argentina	1904	Laurie Island
Palmer Station	US	1968	Anvers Island
Petrel Base*	Argentina	1952	Dundee Island
Primavera Base*	Argentina	1977	Graham Land
Princess Elizabeth Antarctica	Belgium	2007	Queen Maud Land
Profesor Julio Escudero Base	Chile	1994	King George Island
Progress Station*	Russia	1988	Prydz Bay
Rothera Research Station	UK	1975	Adelaide Island
Russkaya Station*	Russia	1980	Marie Byrd Land
San Martin Base	Argentina	1951	Barry Island
South African National Antarctic Expedition (SANAE IV)	South Africa	1962	Vesleskarvet, Queen Maud Land
St. Kliment Ohridski* Base	Bulgaria	1988	Livingston Island
Scott Base	New Zealand	1957	Ross Island
Showa Station	Japan	1957	East Ongul Island
Signy Research* Station	UK	1947	Signy Island
Svea Research* Station	Sweden	1988	Queen Maud Land
Taishan Station*	China	2014	Princess Elizabeth Land
Tor Station*	Norway	1993	Queen Maud Land
Troll Station	Norway	1990	Queen Maud Land
WAIS Divide Camp*	US	2005	West Antarctic Ice Sheet
Wasa Research* Station	Sweden	1989	Queen Maud Land
Vanda Station	New Zealand	1967 (closed 1995)	Lake Vanda, Victoria Land
Vernadsky Research Base	Ukraine	1994	Galindez Island
Vostok Station	Russia	1957	Antarctic Ice Sheet
Zhongshan (Sun Yat Sen) Station	China	1989	Larsemann Hills (Prydz Bay)

*indicates open only during astral summer

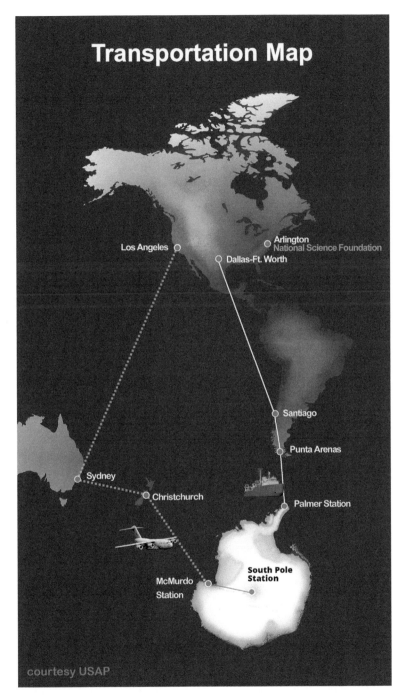

Fig. 1.17. Transportation to US facilities in Antarctica goes through Punta Arenas by sea and through Christchurch by air (Map courtesy USAP)

2

The Mountain and Its Madness

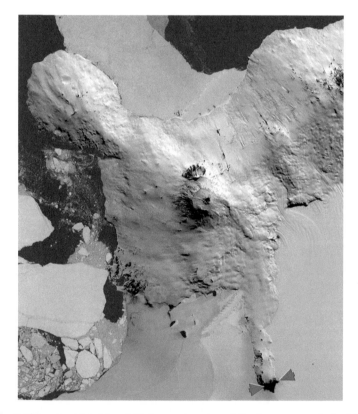

Fig. 2.1. 3,794-meter high Mount Erebus—at center of the image—overlooks McMurdo Station and Scott Base (left and right arrows, respectively) on Ross Island. Mount Terror is to the right (NASA/GSFC/METI/ERSDAC/JAROS and US/Japan ASTER Science Team)

© Springer International Publishing AG, part of Springer Nature 2019
M. Carroll, R. Lopes, *Antarctica: Earth's Own Ice World*, Springer Praxis Books,
https://doi.org/10.1007/978-3-319-74624-1_2

Erebus by the Numbers

Elevation: 3794 meters
Location: Ross Island, Antarctica
First ascent: 1908 (by members of the Shackleton Expedition)
Last major eruption: 2011
Crater dimensions: 500 x 600 meter diameter
Lava lake dimensions: 160 meter diameter

Early Exploration

The lure of Erebus is legendary. One can only imagine early explorers looking up at the majestic mountain from the Ross Ice Shelf and wondering how it could be climbed. What would they find at the top? How could it be summited, given the elevation (3,794 m or 12,448 feet) and the hostile conditions? In 1908, members of Sir Ernest Shackleton's expedition party decided to try and, amazingly, succeeded. The brave mountaineers included Tannatt William Edgeworth David, a 49-year old Welshman and Professor of Geology at Sydney University. The idea of seeing an active Antarctic volcano must have been quite irresistible to the geologist, and he led the climbing party despite being the oldest in the group. Accompanying him was another scientist and Australian resident, 25-year old physicist Douglas Mawson. Mawson definitely caught the Antarctica "bug," as he came back to lead the *Australasian Antarctic Expedition* in 1911–14. He was rewarded with a knighthood in 1914. Another member was the Scottish Dr. Alister Forbes McKay. At 30 years old, he was Assistant Surgeon and, sadly, was lost and presumed dead in 1914 on *Vilhjalmur Stefansson's Canadian Arctic Expedition*. Surgeon Dr Eric Marshall, then 28, was also a skilled cartographer who later served in World War I. He was honored with a CBE (Commander of the Most Excellent Order of the British Empire) and lived to the ripe age of 84. Lieutenant Jameson Boyd Adams, 27 years old and second in command to Shackleton, was a trained meteorologist who went on to serve bravely in the two World Wars. The last in the climbing party was 20-year-old Sir Phillip Brocklehurst who served as Assistant Geologist, but he did not reach the summit due to frostbitten feet that would eventually necessitate the amputation of a big toe. Brocklehurst also served with distinction in both World Wars. He died in 1975, the last of the climbing party.

The climbing team set out on March 5, reaching the summit five strenuous days later. In the words of geologist Edgeworth David:

We stood on the verge of a vast abyss and at first could see neither to the bottom nor across it on account of the huge mass of steam filling the crater and soaring aloft in a column 500 to 1000 foot high. After a continuous loud hissing sound, lasting for some minutes, there would come from below a big dull boom, and immediately great globular masses of steam would rush upwards to swell the volume of the snow-white cloud which ever sways over the crater. This phenomenon recurred at intervals during the whole of our stay at the crater. Meanwhile, the air around us was extremely redolent

Fig. 2.2. Photograph taken by Sir Douglas Mawson at the Erebus crater rim on March 10, 1908 during the first ascent of the mountain. Mawson noted that measurements "made the depth 900 feet, and the greatest width about half a mile. There were evidently three well-like openings at the bottom of the caldron, and it was from these that the steam explosions proceeded." (Image provided by the Royal Geographic Society, United Kingdom)

of burning sulphur. Presently a pleasant northerly breeze fanned away the steam cloud, and at once the whole crater stood revealed to us in all its vast extent and depth.[1]

---

[1] Shackleton, 1909, The Heart of the Antarctic: being the story of the British Antarctic Expedition 1907–1909. W. Heinemann, London.

The climb of Mount Erebus was full of firsts. It provided valuable geological observations of the volcano. It also served to familiarize the men with their equipment and sledging gear. When by October 1908 the weather improved sufficiently to resume their ambitious exploration efforts, Shackleton and his party of four traveled onwards to the South Geographic Pole, while Edgeworth David led Douglas Mawson and Alistair Mackay for the South Magnetic Pole in what became known as the Northern Sledging Party. It was an epic 122-day journey totaling 2,030 kilometers, during which the men carried more than half a ton of equipment and supplies. It remained the longest unsupported man-hauling journey in history until the record was broken in the 1980s. The party not only reached the South Magnetic Pole but gave the most accurate fix on the location yet.

The next ascent of Erebus happened in 1912 during Scott's last expedition, and it was led by geologist Raymond Priestly, who would later collaborate with Edgeworth David on geological descriptions of the volcano. Four out of Priestly's party of six reached the summit, including Tryggve Gran, who was the Scott expedition's ski expert from Norway. He described being at the edge of the crater and hearing a gurgling sound: "…before I had realised what was happening I was enveloped in a choking vapour. The steam cloud had evidently much increased by the eruption, and in it I could see blocks of pumiceous lava, in shape like the halves of volcanic bombs and with bunches of long drawn-out hair-like shreds of glass in their interior. The snow around me was covered with rock dust and the smoke was yellow with sulphur and disagreeable in the extreme."

We rightly marvel at what the early Antarctica explorers achieved, especially as traveling to and staying on the mountain is still challenging. Helicopters and snowmobiles make the ascent easy, but not for one moment can we forget the potential dangers. The weather can turn quickly. Helicopters have an impressive safety record, but they travel over treacherous terrain, and survival gear must be onboard at all times. Flying over the vast white flanks of the mountain in places cut by crevasses is a reminder of how close to danger one is at all times. Snowmobile accidents do happen, and evacuating people from the mountain could take days if the weather is bad. The volcano itself is always a potential threat. Erebus is active, mostly quietly, housing a bubbling lava lake in its crater. However, larger explosions have occurred. Like all volcanoes, Erebus has to be treated with great respect.

The Volcano and its Setting

Mount Erebus is the most active volcano in Antarctica and one of the largest active volcanoes on Earth. Other volcanoes in the frozen continent have seldom erupted in historic times. As far as we can tell, Erebus has been erupting constantly since it was discovered in 1841, though the level of activity varies. Most of the time, the lava lake in the crater quietly churns, but on some occasions, Strombolian explosions from the crater have sent volcanic bombs (lava fragments) flying hundreds of meters away. In 1984, violent explosions sent bombs nearly 3 km away from the crater!

The upper part of Mt. Erebus is a stratovolcano and the lower part is a shield volcano. A stratovolcano is made up of numerous layers of lavas and ash and other fragments thrown out by explosive eruptions, while shield volcanoes pile up from many lava flows. Mount Fuji in Japan is a classic example of a serene-looking stratovolcano. Shield volcanoes are common in Iceland and Hawaii – the name "shield volcano" actually comes from

Iceland because their profiles look similar to warrior's shields on the ground. Mauna Loa in Hawaii is a typical example of a shield volcano.

Unlike many other volcanoes on Earth, Erebus is not located atop a tectonic plate boundary (regions where the Earth's plates are spreading apart or one plate is subsiding under the other). Erebus is what is known as an intra-plate volcano that lies in a rift where the Earth's crust has been thinned, such as in the East Africa Rift Valley or in the western USA (where Yellowstone is located). Erebus lies on the West Antarctica Rift system, an active rift valley located between east and west Antarctica. Furthermore, Erebus sits atop a mantle plume known as the Erebus hotspot, which is responsible for the relatively high level of volcanic activity on Ross Island. Both Mount Bird at the northern end of Ross Island and Mount Terror at its eastern end are volcanoes as well, but older. Potassium-argon dating of their rocks indicates that they have not erupted for at least a million years. In contrast, Erebus rocks have been dated from just over a million years old for the lower shield volcano to less than 100,000 years old for the upper stratocone. There are many other volcanoes in Antarctica but none as active as Erebus. Recent studies have suggested that numerous volcanoes lie under the ice, prompting concern that renewed volcanic activity could cause massive melting.

Fig. 2.3. The harsh conditions on Erebus have not deterred researchers from studying the volcano (Photo by the authors)

The remoteness and harsh conditions of Antarctica have not deterred scientists from studying Erebus. Prof. Clive Oppenhneimer from Cambridge University in the United Kingdom has been studying the volcano for many years. He considers Erebus to be a great natural laboratory for investigating volcanic and magmatic processes because of its persistent low-level activity that makes observations from the crater rim relatively easy and safe.

We met Clive, along with long-time Erebus researcher Prof. Phil Kyle from the New Mexico Institute for Mining and Technology, at McMurdo. Kyle, a geochemist, made his first trip to Erebus in 1969, coming back so many times that he became known as "Mr. Erebus" due to his pioneering work. When we met Clive and Phil, they were on their way home, while we were on our way up to the volcano. It was, sadly, the last year these two expected to visit Erebus, their research project having ended. When we got to the Lower Erebus Hut, we were witness to the dismantling of some of the equipment that Kyle, Oppenheimer, and their teams had used for monitoring Erebus. Their long collaboration involved studies not only of the activity of the lava lake in the summit crater but also of the gases and volcanic rocks, as well as seismic studies and ground deformation measurements that can reveal if magma is pushing up from below. The volcano is monitored by seismometers, a video camera, infrasound sensors to record the sound of explosions in the lava lake, and other sensors to examine ground deformation and weather. It was fun for us while still at McMurdo to see images from the camera mounted on the rim. The camera monitors explosions that sometimes nearly wipe out the camera itself! Although some of the instrumentation was being taken down, other studies continue. The wealth of data collected over many years has enabled a much deeper understanding of the volcano and generated at least 25 Ph.D. theses. Clive, whose specialty is gas geochemistry, used a Fourier Transform infrared spectrometer mounted on the crater rim to measure seven different gas species and their variations in the scale of seconds. Variations in gases can be an important precursor sign to increased volcanic activity, and the gases themselves give information on the magma chamber. The extremely dry atmosphere of Antarctica is very good for these types of measurements.

Erebus was for some time the best monitored lava lake on the planet, despite its remote and hard-to-access location. Installing equipment in the conditions at the top is hard, sometimes dangerous work. As we were able to witness, taking down equipment was also no easy task. Instruments and power supplies used on Erebus need to be very sturdy and reliable, able to survive the dark winter, extreme cold, howling winds, and storms. Ice accumulates on everything, high winds can damage instruments, and the several months of darkness when there is no solar power to recharge batteries make it challenging or downright impossible to keep instruments running year-round.

Equipment left on Erebus, however, does have the advantage of lacking thieves and vandals in the vicinity – in fact, in the whole continent.

Lava Lake Activity

Erebus has the distinction of being the southernmost active volcano in the world and also the only one to contain a phonolitic lava lake. Phonolite is a fairly rare type of lava rock whose name comes from the Greek word meaning "sound stone" because of the ringing sound it can produce when hit with a hammer. Lava lakes themselves are rare, at least on Earth. The Erebus crater contains one of six or so persistent terrestrial lava lakes, the others being on the volcanoes Erta Ale (Ethiopia), Nyiragongo (Congo), Kilauea (Hawaii), Maum and Benbow (Ambrym Island, Vanuatu). A few other volcanoes have intermittent lava lakes such as Masaya in Nicaragua and Villarica in Chile. At present time, the

Halemamau lava lake on Kilauea in the Big Island of Hawaii is the largest lava lake on Earth. It is small in comparison with those on Jupiter's moon Io.

Lava lakes exhibit quite benign volcanic activity, given that they bubble away within the confinement of a crater. Most lava lakes contain basaltic lava, but Erebus' phonolite is a lot more viscous than basalt. The Erebus lake doesn't churn vigorously like the lake inside Marum crater in Ambrym, though it is still mesmerizing to watch. The behavior of the Erebus lake may be different from that of other lava lakes because of its magma composition. Although the viscosity of the lava in the Erebus lake has not been directly measured, it is expected that it will be about 100 times higher than that of basaltic composition lava lakes such as Kilauea and Erta Ale. The Erebus lava lake moves like treacle, while the Marum lake at Ambrym, which Rosaly Lopes has been lucky enough to see in person, churns so vigorously that it looks like a bubbling cauldron of tomato soup.

Although lava lakes are rare, they are important in volcanology. Lava lakes are the exposed tops of magma systems, and therefore provide volcanologists the opportunity to directly observe the dynamics of magma transport and to infer degassing processes or, as Phil Kyle puts it, "the guts of the volcano." The Erebus lava lake has been persistently active since at least 1972, though it is thought to have existed for much longer, perhaps a century or more.

Monitoring the gases coming out of the volcano (or, as Kyle has put it, the volcano's breath) has revealed an interesting pattern. Clive Oppenheimer first noticed that the total amount of gas rises and falls in a reliable 10-minute long cycle. The gas contains about equal parts water vapor and carbon dioxide (CO_2) with minor amounts of sulfur dioxide (SO_2) and hydrogen chloride (H_2S), the latter being responsible for the "rotten egg" smell that most people hate but volcanologists tend to be peculiarly fond of. The 10-minute pattern, initially observed during one season, might have been temporary, had not short data sets from multiple seasons confirmed the pattern, varying from 5 to 18 minutes. Their colleague Nial Peters (who was one of our campmates on Erebus) was able to show that the lava in the lake rises and falls by about 6 to 10 feet (2-3 m) and lava cooling on the lake surface cracks and flows outwards, keeping time with the gas cycle. The interpretation is that as fresh magma comes up into the lake, the lava at the top rises and spreads out, releasing gases. Nial compared it to a lava lamp, though stressing that the details are quite different, as batches of fresh magma flow as blobs in a lava lamp and not as a continuous supply. We can imagine the blobs in an underground pipe that feeds the lava lake from a deeply buried magma chamber, with lava moving up and down the pipe. Nial noted that the behavior of Erebus has stayed remarkably constant for years, which is unusual for volcanoes.

The gas studies also revealed that Erebus is the largest point source of numerous gas and aerosol species to the Antarctic troposphere, some of which (including sulfur, halogen and nitrogen compounds) are of environmental significance. An important scientific result shows that Erebus is a significant source of nitrogen dioxide (NO_2). This gas, although not a primary volcanic gas, is of particular interest because of its important role in tropospheric ozone chemistry. It is likely that NO_2 is formed by the interaction of atmospheric nitrogen with the hot surface of the lava lake and that Erebus is the main point source for NO_2 (and very likely other reactive nitrogen oxides) in the Antarctic troposphere. This is a good example of how the study of volcanoes has important implications for the environment.

The researchers' high time resolution infrared spectroscopy measurements of lava lake gas emissions also gave important information on the dynamics and geometry of the plumbing system under the volcano. Volcanologists use the term "plumbing" to refer to the system of conduits that bring magma to the surface. The gas measurements indicated that the bulk of the gas emitted by Erebus is sourced two kilometers below the lava lake yet is in chemical equilibrium with magma at the surface. They also find clear evidence that only part of the deep magma that yields carbon dioxide rises to shallower levels in the volcano's plumbing system.

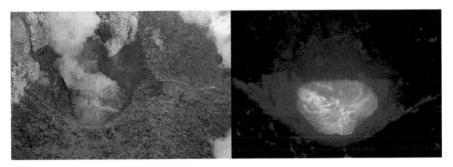

Fig. 2.4. Visible and infrared images of the Erebus lava lake taken on December 15, 2013 by Nial Peters using a thermal infrared (IR) camera situated on the northern rim of Erebus's main crater (Images copyright and courtesy Nial Peters)

Ice Caves

Heat and gases emanating from inside Erebus form Fumarolic ice caves, known as FICs, and the amazing towering structures above them. These may be important analogues for phenomena on other bodies such as Enceladus and Europa. Aaron Curtis, who is now Rosaly's colleague at the Jet Propulsion Laboratory, did his Ph.D. research under the mentorship of Phil Kyle on the Fumarlic ice caves of Erebus. Very little was known about the caves before then. As Aaron pointed out in his thesis, his first job was to do a literature search, where he found that published work on the FICs consisted of only 13 pages.

The Fumarolic ice caves are geothermal features that Aaron used as a way to quantify and understand flank degassing. Although there are FICs on other volcanoes such as Rainier in the Cascades, there are a lot of them on Erebus – over 100. The caves are networks of passages melted into the base of the snowpack where heat and warm gases come up to the ice-rock interface. Except at the entrance to the caves, permafrost and ice cover typically seal off the bedrock surface to gas release. Above many FICs entrances we see spectacular hollow, conical towers of ice up to 15 meters tall, known as Fumarolic ice towers (FITs).

Degassing from the flanks is a common phenomenon on volcanoes, but nowhere is this seen more dramatically than at Erebus. The towers were the first geothermal features observed on Erebus. When the humidity is relatively high (above 15%), steam plumes from the towers can be seen for miles. In fact, they were noticed during the first ascent of

Erebus. In 1909, Edgeworth David and his colleague R.E. Priestley wrote about the towers, which they called "ice fumaroles," and mentioned that these along with the large feldspars were "two features in the geology of Erebus which are specially distinctive."[2] They couldn't have known that there are miles of horizontally developed passages underneath, some up to 10 m wide and 8 m tall. Aaron Curtis writes in his thesis, "It is unclear at what point researchers became aware of the extensive cave passage development, and why those passages did not become an immediate scientific target. The cumulative difficulties of working in a frontier (the caves) of a frontier (the Erebus caldera) of a frontier (Antarctica) were presumably a barrier."

The caves are not static, as Erebus continues degassing. Aaron's work showed that the Fumarolic ice caves change on the scale of tens of centimeters annually, and that the topography above the caves also changes due to the enlargement of chambers through melting. Aaron's work opened up a lot of interest in the Erebus FICs, and he has provided future researchers with a substantial amount of data, including the Erebus Cave and Fumarole Database (http://erebuscaves.nmt.edu).

Fig. 2.5. Cutaway view of a typical ice cave/tower formation on Mt. Erebus (Digital rendering by Michael Carroll, based on diagram by Curtis and Twombley)

Erebus Crystals

The large feldspar crystals first mentioned by Edgeworth David and Priestley in 1909 are popularly known as "Erebus crystals." They are the most coveted of souvenirs in Antarctica, although collecting them is now against the rules even though, as one researcher pointed

---

[2] David, T. W. E., and R. E. Priestley (1909), Geological observations in Antarctica by the British Antarctic Expedition, 1907-1909, in The heart of the Antarctic, vol. 2, pp. 276–331.

out, the volcano keeps making more. The feldspar crystals are unusual because they are so large compared with most found elsewhere. The crystals are rich in sodium, potassium, and aluminum silicate. Only two other volcanoes in the world are known to produce similarly large crystals, and they are located very far from Antarctica: Mount Kenya and Mount Kilimanjaro in Africa, though their chemical compositions are slightly different.

Edgeworth and Priestley wrote that as they approached the Erebus crater, "the covering of snow became thinner until it almost entirely disappeared, being replaced by a surface formed of crystals of anorthoclase feldspar from half an inch to four inches in length."[3] This is a sight that still greets climbers as they approach the crater, and we were delighted to see so many, having spotted only a couple near the Lower Erebus Hut. Why these crystals grow so large is still not well understood, though it is clear that they grow slowly in the magma. Work by Yves Moussallam, Nelia Dunbar, and colleagues (in cooperation with Mt Erebus Volcano Observatory) demonstrates that the crystals grow large in an onion-like fashion, adding outer layers as they rise and sink in the convection of the lake. The crystals are ejected out of the lava lake during explosions, encased inside glassy volcanic bombs. Volcanic bombs are magma fragments that come out of a volcano during explosions; by definition the fragments are larger than 6 cm. Volcanic bombs fall to the ground and splatter without exploding. The Erebus bombs have glassy crusts, meaning that their outer layer cooled very quickly without crystallizing. This glassy crust is easily breakable and weathers away, exposing the feldspar crystals inside. The result is that the upper slopes of Erebus are left covered by the crystals, usually loose, though sometimes one finds bombs with the crystals sticking out from the sides.

Lava Lakes on Io

Jupiter's moon Io is the most volcanically active body in the Solar System and is also the only place outside the Earth where eruptions of molten rock (magma) are still going on. We know that volcanism happened on other bodies such as Venus, Mars, Mercury and the Earth's Moon, but at present, those bodies exhibit no ongoing activity, though there have been some tantalizing suggestions that Venus may still have some active volcanoes or at least degassing. Io, although only the size of the Earth's moon and thus expected to have cooled long ago, is still volcanically active thanks to its orbit between Jupiter and the other Galilean satellites Europa, Ganymede, and Callisto. Volcanism on Io occurs thanks to tidal heating that keeps the interior of the moon hot, allowing magma to remain molten. Unlike Earth, Io has no plate tectonics.

The first spacecraft to reveal Io's volcanoes was Voyager 1 in 1979, followed by Voyager 2 a few months after. The Voyager flybys showed Io's colorful surface, dotted with volcanic-looking features. When volcanic plumes were spotted, it was clear that this little moon was beyond special, with volcanoes erupting at several locations. The exploration of Io continued with the Galileo spacecraft, which orbited Jupiter starting in 1995 and made numerous observations of Io's volcanoes. The surface of Io is covered by deposits of sulfur

---

[3] T. W. E. David and R. E. Priestley *British Antarctic Expedition*, 1907-9. Reports on the Scientific Investigations, 1914, vol. 1, p. 213 and pl. 65.

Fig. 2.6. This lava outcrop is filled with the infamous Erebus feldspar crystals (Photo by the authors)

and sulfur dioxide with colors ranging from dark reds to oranges, yellows, and whites. The active or recently active lavas are dark, and as they cool, they become mantled by sulfur and sulfur dioxide deposits. Volcanic plumes rise up to 500 km above the surface, with sulfurous deposits forming gigantic rings as the material falls down. The Galileo space-craft obtained images in the visible and infrared over several years, some at high spatial resolution, which allowed us to study eruption styles. Explosive volcanism, characterized by large plumes and high thermal emission, is the most spectacular. These eruptions can make noticeable changes on Io's surface, producing large plume deposits. Then there are lava flows, which sometimes create smaller plumes at their distal ends as they move across the surface. However, the most common type of volcanic activity on Io is volcanism confined within a caldera-like depression. The depressions are called paterae (meaning shallow depression) on Io and other bodies rather than calderas because we cannot be sure that they were formed the same way as terrestrial calderas. However, images of Io's pat-erae in both the visible and the near-infrared suggest they are morphologically similar to terrestrial calderas, with floors that in many cases appear to contain lava lakes.

Lava lakes on Io were discovered from the thermal signatures of high resolution images from the Near-Infrared Mapping Spectrometer (NIMS),[4] which revealed hot margins at the edges of several paterae, similar to terrestrial lava lakes where the crust is broken up as

---

[4]Lopes, R.M.C., et al. 2004. "Lava Lakes on Io. Observations of Io's Volcanic Activity from Galileo during the 2001 Flybys." *Icarus*, 169/1, pp. 140-174.

the crusted-over surface of the lava lake hits the crater walls. This is relatively common on lava lakes on Earth, and to understand it we can make an analogy of scum on top of a pond. As the water sloshes around, the scum hits the margins and is broken up.

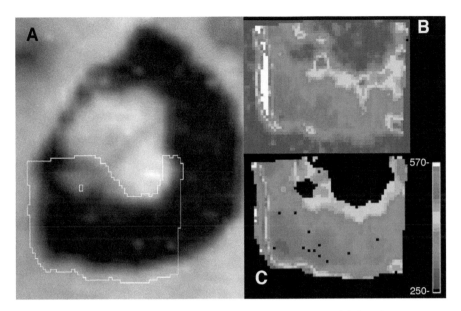

Fig. 2.7. Loki patera on Io imaged by Galileo's camera SSI (*A*) and infrared spectrometer NIMS at 2.5 microns (*B*). A temperature map derived from NIMS images is in image *C*. Note the hotter edges (represented by red and white colors), characteristic of a lava lake (NASA/JPL/Rosaly Lopes)

Images of Io's paterae taken by the Galileo camera also showed evidence of lava lakes. In the case of one patera, Tupan, thin, dark margins can be seen where the cooled crust of the lava lake meets the patera walls. In general, the darkest features on Io at visible wavelengths are hot, active lavas. These lava lakes are much larger than those on Earth – for example, the Loki patera, which is thought to contain a lava lake, is close to 200 km across. In comparison, the largest Earth lava lake at present is located inside the Halema'uma'u pit in the island of Hawaii. Situated atop Kilauea volcano, Halema'uma'u became active again in 2008 after decades of repose. Halema'uma'u is now roughly 195 m by 255 m (about 640 x 840 ft) in size and may grow even more, but it is small compared to the gigantic lava lakes on Io.

More Weird Lakes?

Lake-like features have been spotted on other worlds, but actual lava lakes akin to that of Erebus are hard to come by. For example, high resolution imaging from the MESSENGER orbiter revealed extensive volcanic remnants on the planet Mercury, but MESSENGER's

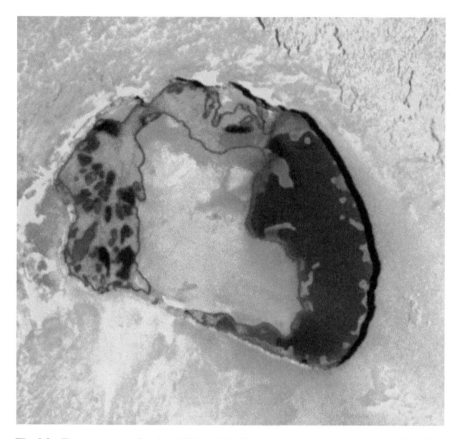

Fig. 2.8. Tupan patera on Io, about 75 km (47 miles) across, is thought to contain a lava lake surrounding a cold central island on which sulfurous materials (yellow) have condensed. Dark areas correspond to hot materials. Dark (hot) margins located next to the patera wall on the western (left) side and against the central island are thought to be due to the cooled surface of the lava lake breaking up against the walls. Hot material is still underneath, causing the cold crust to move around and hit the walls (NASA/JPL)

six months of close orbital surveillance show that the lava flows are different from those on Earth. No volcanic vents, lakes, or cones are visible at most of the extensive lava flows, and experts suggest that the sources of lava are entombed beneath the flows.

When the lava flowed out of cracks during the planet's infancy some 3.5 billion to 4 billion years ago, it filled craters more than a mile (1.6 km) deep. Many locations have been invaded by small, shallow, irregularly shaped depressions called hollows, which may be sites of volcanic gas venting. Yet, the closest planet to the Sun has no evidence of lava lakes.

Venus, a world covered by volcanically related geology, seems a better place to search, and the European Space Agency's Venus Express orbiter has done just that. The venerable craft orbited Venus from 2006 to 2014. During that interval, sensors charted 'hot spots' at several sites. The apparently heated patches of real estate were short-lived, with a thermal signature that would be expected at an active volcano. Some isolated features heated up over the course of a few days and then quickly cooled. Lava flows quenched by Venus' high-pressure atmosphere would fit the temperature model nicely; however, there is insufficient data to show that this is really the case.

Lava lakes may have occurred on Venus in the more distant past. Caldera-like features that might once have contained lava lakes are abundant on Venus. According to researchers Pete Mouginis-Mark and Scott Rowland from the University of Hawaii, the smooth floors of some calderas such as Sacajawea patera (Fig. 2.9.) likely once contained lava lakes.

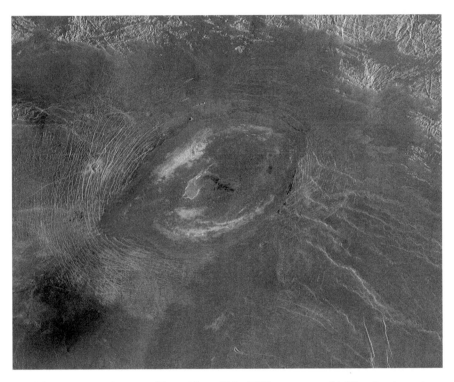

Fig. 2.9. Sacajawea Patera on Venus (about 120 x 250 km across and 1-2 km deep) may have contained a lava lake. Synthetic Aperture Radar image taken by the Magellan spacecraft (NASA/JPL)

The Earth's other planetary neighbor, Mars, is no slouch when it comes to volcanoes, and several Martian volcanoes share structural similarities with terrestrial volcanic craters and calderas. Pete Mouginis-Mark and Scott Rowland proposed that some of the Martian calderas also contained lava lakes in the past. One clue comes from the smooth surface of calderas, such as the nested calderas atop Olympus Mons. They used topographic data from the Mars Observer Laser Altimeter (MOLA) to show a prominent downsag on the floor of the largest Olympus Mons caldera, suggesting a past lava lake. But to confirm the existence of ancient lava lakes on Mars, there is still work to be done. Higher resolution topographic data can reveal surfaces within calderas that have settled into a stable, equipotential surface. Even higher resolution imaging, on the order of less than 10 m/ pixel, can help to discern between lava flows and lava lakes by revealing the morphology of precise flow margins in detail. Researchers are also searching for polygonal cracks that typically form on the surface of cooling lava lakes.

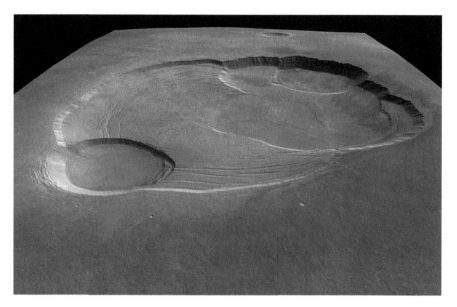

Fig. 2.10. The Olympus Mons summit caldera complex. The largest is about 60 km across. The smooth surfaces and downsag suggest it was once filled by a lava lake (NASA/JPL)

Ice moons like Europa and Ganymede (at Jupiter), Enceladus and Titan (at Saturn), and Miranda and Ariel (at Uranus) exhibit features that may be attributed to cryovolcanic collapse or flow. But none of them appear to have ever been long-lived features that operated like terrestrial lava lakes. We have caught cryovolcanism in action at Enceladus with Cassini and at Neptune's largest moon Triton with Voyager. Several Voyager images show columns of dark material drifting nearly vertically for about 8 km. At that altitude, the plumes apparently encounter a jet stream that carries the dark material for dozens of kilometers, draping it in long blankets across the surface of the ice moon. But this bizarre form of cryovolcanism is not the only type seen on Triton.

Of the 30% of territory imaged in detail by Voyager, several cryovolcanic calderas have tentatively been identified. Smooth-floored depressions, some with frozen wave-forms, are constrained by a set of circular cliffs. Complex, pitted centers of the flat regions may be collapsed lava tubes. Gentle domes rise from various locations across the floors, indicating multiple stages of cryovolcanic activity, and the general form of the features is that of a frozen pond.

Within Triton's frozen nitrogen ices, dark, lake-like features have settled into the landscape. Called guttae, the lobate blotches span some 100 to 200 km across. Each is surrounded by a bright halo 20 to 30 km wide. The smooth surface and lobate forms of the guttae suggest that viscous material has been forced onto the surface, where it piles up at least a few tens of meters thick. Several explanations have been put forward for the origin of these mysterious features, including low-viscosity cryolava flows filling a depression, reheating of surface ices, or sites of venting gases. But the guttae display none of the characteristics of a long-lived lava lake. One other Solar System member might: Pluto.

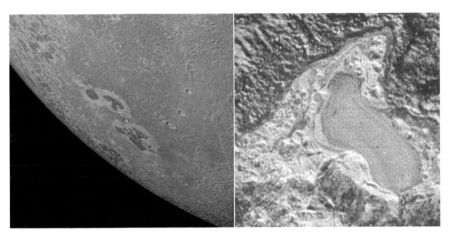

Fig. 2.11.   (*Left*) Voyager 2 image of Triton. The field of view is about 1,300 km across. Three irregular dark areas (guttae), surrounded by brighter material, dominate the lower image (NASA/JPL); (*Right*) the mysterious ponded feature Alcyonia Lacus rests in the mountains bordering Pluto's northern Sputnik Planitia (New Horizons image courtesy NASA/JPL/JHUAPL/SwRI)

Two large structures on Pluto have been tentatively identified as towering cryovolcanic constructs, thanks to the images returned by the New Horizons spacecraft in 2015 (see Chapter 7). Wright Mons and Piccard Mons resemble shield volcanoes. Concentric ledges and fractures crown their summits. Their gentle, undulating slopes fan out as wide as 150 km (90 miles), and their summits rise some 4 km (2.5 miles) high. If these are in fact volcanoes, they are the largest beyond Mars. Due to the low sun angle during encounter imaging, the interiors of the summit depressions were in shadow. It is possible that they may have had active cryolava lakes at some time in their pasts, but any true lakes have yet to be confirmed on the ice dwarf world. It seems unlikely that any are active today.

So, it appears that Erebus-like cryolava lakes do not occur today among the ice moons and dwarf planets of the outer Solar System. The active lava lakes of Earth along with those on Io may be unique among volcanic features across the modern Solar System. This is one of many reasons that the simmering lake within the crater atop Erebus draws researchers from all corners of the globe and from many disciplines. But getting there is a formidable task.

Fig. 2.12. Artist rendering of Wright Mons based on the summit crater of Erebus (Painting by Michael Carroll)

3

First Step: Getting There

For US citizens, the journey to McMurdo Station, Antarctica begins not at a major airport but in Washington, D.C., home of the National Science Foundation. More specifically, our project began with a proposal to the NSF, which gave grants to roughly 500 proposals in 2016. Only four of those were granted to the Artists and Writers Program.

Our first proposal was rejected primarily because our Principal Investigator (PI), Rosaly Lopes, spent more of her time being a scientist than being an artist/writer. Despite the fact that Lopes had authored several books, NSF felt the proposal would be stronger if the designated PI was Carroll, who was a full-time artist and writer.

Our second submission met with success. But our project, which we called "Alien Landscapes," was to hit one more snag: logistics. As we prepared to instigate a 2015 expedition, resources in Antarctica became overwhelmed by project scheduling and weather. Due to a lack of helicopter availability, we had to delay for one more year. Finally, in 2016, the NSF's United States Antarctic Programs pulled the trigger on our proposal. It was finally time to go.

The Proposal

The subhead of our Alien Landscapes proposal explained that we would be "Melding art and science to relate the beauty and significance of Antarctica's landscapes to the unseen vistas of other worlds." Our application totaled 15 pages and included photos and charts showing planetary analogs, planned activities, and sample paintings. The heart of our project was best summed up as: "Ice and lava have interacted to create unique landscapes not only on Earth but also on other planets." Our proposal showed comparisons of Icelandic valleys to fractures on Triton, terrestrial lava lakes to those on Jupiter's moon Io, and Earthly volcanoes to those on Mars.

We included reviews of our past books and lists of articles we had written, both for technical journals and popular science magazines. The document surveyed past work, including Lopes' exploration of volcanoes worldwide and Carroll's field experience with

© Springer International Publishing AG, part of Springer Nature 2019
M. Carroll, R. Lopes, *Antarctica: Earth's Own Ice World*, Springer Praxis Books,
https://doi.org/10.1007/978-3-319-74624-1_3

various scientific expeditions, including a 13-week US Geological Survey's analysis of the Bering Glacier, where he did on-site paintings and reporting for magazines and for the USGS artists program. Our proposal also included supporting materials: copies of our books, Curricula Vitae, letters from editors in the publishing industry, and letters of recommendation from scientists.

After we heard the good news that our proposal had been accepted, we had to work on another document, the Research Support Plan, with our implementer Michael Lucibella. An important part of the process involved evaluating the potential impact our work would have on Antarctic travel resources. As we were requesting access to Mt Erebus and the Lower Erebus Hut field camp, our itinerary was more complex than most expeditions. "We request helicopter support to transport us to an acclimatization camp on the Fang Glacier and subsequently to the Lower Erebus Hut. We request access to the crater rim, Tower Ridge, and non−pristine/non−technical ice caves if possible (though we do not need to enter the caves, but rather look at their local setting). We request access to the LEH skidoo pool when not in use for science to access field sites beyond walking distance. We also request one FST (Field Safety Trained) member to accompany us to field sites, as we are unfamiliar with the sites and conditions. We understand that weather can be a problem, and will depend on local personnel's advice for all logistics."

The NSF requires answers to a host of questions pertaining to the Antarctic environment and McMurdo's infrastructure as well. A sampling of questions includes:

Does your project require drilling or coring of ice, rock, or sediment?
Does your project require performing lab work or any use of hazardous materials in the field? Generating any hazardous wastes in a lab or field location?
Will your activities involve evacuation of soil or snow?
Excluding the emissions from the combustion of fossil fuels, will the proposed activities result in any release into the Antarctic environment, including irretrievable science equipment, hazardous materials, wastewater, etc?
Do you require the construction or use of field camp structures?
Do you require the fabrication of Special Shipping Crates?
Do you require the Modification of an existing Lab Space?
Will you be requiring the use of explosives?

Finally, we outlined the expected significance of our work "…to bring the beauty and the universal significance of Antarctica to the general public." We asserted that our lectures, book and articles on the subject would be resources for both art and science students, as well as for the general public interested in Antarctica, art, geology, and astronomy. Our overarching aim, the proposal read, was to "show that art, science, and exploration go hand-in-hand, particularly in places like Antarctica, where artists and scientists have to be explorers and work together."

After our several years of proposal toil, our project was in a form acceptable to the wise counselors at the National Science Foundation. We eventually received the all-important Support Information Package (SIP), which spells out the field plan and what support NSF will provide. But there were more hurdles to jump over, and that course would be set by the University of Texas Medical Branch (UTMB), the gatekeepers of Antarctica's physical health and wellbeing.

Pre-Qualification

The UTMB manages the health of all three United States Antarctic Programs stations: McMurdo, Palmer Station, and the Amundsen/Scott South Pole Station. The 84-acre center operates out of the University of Texas. Its Center for Polar Medical Operations oversees medical personnel in Antarctica. Each facility on the seventh continent has a physician, and McMurdo has the equivalent of a level four urgent care center equipped with radiology and lab areas. McMurdo also supports either an Emergency Medical Technician or Physician's Assistant during the height of the season, as well as a medical laboratory technician. UTMB is responsible for the resupply of all three primary stations with medical supplies and medications. They also carry out real-time medical support to remote medical facilities through the use of telemedicine, video conferencing, radio and other techniques. Additionally, the center services the many seasonal field camps, along with two research ships that operate throughout the year.

Fig. 3.1. *(Left)* McMurdo's medical center provides a level four urgent care site. The center is ready for a host of conceivable crises, as evidenced by this sign *(Right)* (Photos by the authors)

UTMB vets the estimated 3,000 people who travel to—and live at—the US Antarctic centers year-round. This includes the grantees, all of whom must be medically prequalified before they are allowed to venture south. The PQ process is a convoluted one, and rightly so. Medical emergencies on the ice are serious affairs and are avoided or preempted at all costs. Pre-existing conditions can be cause for a prospective worker, researcher or grantee to be disqualified from traveling to Antarctica. The official literature from the Antarctic Support Contract team (ASC Leidos) puts it this way: "The physical qualification (PQ) process administered by ASC seeks to screen out people with conditions that cannot effectively be managed on the Ice or aboard ship." Candidates who do not meet the U S Antarctic Program physical qualifications may apply for a waiver, which is reviewed by NSF. The UTMB's process parallels NASA's and ESA's vetting of astronauts (for similar reasons), and a comparable system will likely be in place for travelers to stations on other worlds (though certainly more rigorous).

As part of pre-qualification, candidates must update all of their immunizations. The PQ process includes laboratory tests such as chest X-rays, blood work, breast exams for both males and females, and even tests for STDs. Candidates are encouraged to schedule their examinations early to ensure that complete information can be provided to the USAP no less than eight weeks prior to their planned departure for Antarctica. In addition to medical forms, other documents are submitted by every candidate. Passing the physical examinations means that the candidate is "PQ'd." Personnel are not ticketed to travel to Antarctica until all required documents have been returned to the UTMB. Ticket information is normally sent two weeks before departure date. Candidates are advised to submit information as early as possible to allow time for resolving problems or re-testing.

Members of other nations' Antarctic program must be approved through their own country's health vetting. They must provide UTMB with their country's clearance documentation if they are going to US facilities. All visitors to US facilities are asked to carry a copy of their medical records with them in case the need arises for US medical services. A resident of a nation that does not have an Antarctic field program will be required to pass the UTMB's PQ screening process.

Fig. 3.2. A gingerbread house incorporated the UTMB as its theme in the annual McMurdo gingerbread house contest, underscoring the important part the UTMB plays in the Antarctic experience (Photo by the authors)

UTMB provides a series of forms that must be filled out by the candidate. These include such topics as "Medical Risks for NSF-Sponsored Personnel Traveling to Antarctica," a five-page medical history, a two-page examination form for the physician, and checklists for required lab tests. A full dental exam is required (with up-to-date full set of X-rays), along with updates to any dental work that the dentist feels is needed before travel.

Between the laboratory tests, letters from physicians and dentists, and requested additional tests, our PQ process took several months. With our acceptance letters in hand, we were at last ready to make final preparations for our expedition. These preparations included much last-minute purchases at various camping stores and travel outlets. Although a set of equipment is provided by USAP, we were counseled by veterans to buy such things as good gloves that fit us well, thermal socks, prescription sunglasses, hats, and the all-important "base layers," or thermal underwear (something one should never scrimp on). Once our polar fashion was complete, we headed for our respective airports.

The USAP booked each of us on separate flights coming in slightly different directions, with Rosaly staging from Pasadena and Mike from Denver. Itineraries took advantage of the cheapest air fares so could not be changed by the passengers. After 32 hours in the air with various layovers in Dallas, Sydney, LA, and Auckland, the two weary travelers met in Christchurch, New Zealand, the final staging arena for all flights to McMurdo.

The Christchurch Experience

We arrived in Christchurch, New Zealand on different flights and were transported to a no-frills but comfortable hotel for which we had to pay ourselves (hotel in Christchurch is not included in the expenses NSF supports for the Artists and Writers program). The next morning, we were taken to the USAP facility to get our cold-weather clothing. We had sent a list ahead of time with sizes and requirements, so we each found two bags waiting for us in the men's and ladies' dressing rooms. It is important to try everything on ahead of time – we had to bring in our own socks and base layers to make sure our boots and outer layers fit properly. The conditions in Antarctica are sufficiently harsh without the additional discomfort of ill-fitting boots or clothing. We each got two coats, known as the "Little Red" (the lighter one, for milder conditions) and the "Big Red" (which we wore most of the time outdoors). If you bring your own outer gear, as most of the mountaineers who work for the program do, it has to be approved by USAP ahead of time. If you are missing something or some of your gear does not fit well, you can request an exchange at windows (screened by curtains) from either dressing room. The gear distribution and storage center is large and impressive, with rows upon rows of cold weather gear.

The people at the center were extremely helpful and stored our suitcases, including everything we might have needed for New Zealand (but not Antarctica). We were limited to 34 kg (75 lbs) of baggage on the flight to McMurdo, and the cold weather gear was

Fig. 3.3. The cold weather gear distribution center in Christchurch (Photo by the authors)

heavy. All luggage had to be weighed prior to our flight and, before we boarded the plane, officials weighed us along with all our cold-weather gear.

After all of this equipment was collected, we had the first of many training lectures about Antarctica conditions and conservation. The lecture included details about our flight, which we hoped would be the following morning. Flights are extremely dependent on weather and many are postponed if the conditions in either Christchurch or McMurdo in particular do not look good. You are also advised that it is not uncommon for a flight to "boomerang." After the flight departs but before getting to "the point of no return" over the water, the pilots contact McMurdo to find out the latest weather conditions, which can change fast in Antarctica. If conditions don't look good, the flight returns to Christchurch. For that reason, we pack a carry-on size "boomerang" overnight (or over several nights) bag, since our luggage will remain secured in the plane.

We were given clear instructions on our pick-up the following morning, including time at the USAP facility to repack as needed, and on the check-in procedures (in our jet-lagged state, we both misheard the kiwi accent as "chicken" instead of "check-in," which was very puzzling). The next steps were computer security training and a last check of our medical forms, including proof of current flu vaccinations.

Computer Security

The United States Antarctic Program provides the front line in cyber security for all US Antarctic facilities. USAP complies with all US federal government security and operational requirements for computing systems by screening all electronic devices, including scientific/research systems, personal computers, laptops, smart phones, tablet devices, and Personal Digital Assistance (PDA) systems. USAP experts screen these devices before grantees are allowed to connect to the USAP's network.

Screening technicians enrolled us in a formal class that presented the importance of information security, antivirus definitions, patches, and vulnerability remediation. All of our Operating System and Software Patches were required to be a version currently supported by the USAP and were updated with the most current patch level of the OS, including the latest security patches. USAP software experts provided our devices with the most current version of antivirus software, configured for auto–updates. All computers in Antarctica must be virus free to connect to the USAP network and maintain the current DAT version as updates are available. Telnet and FTP were prohibited, as they present a high risk to the USAP network. Communication is of paramount importance for safety of personnel and to be able to conduct scientific and technical work.

The International Antarctic Center

Across from the USAP facility in Christchurch, there is a museum called the International Antarctic Centre (note the British spelling) that is considered by many the prime tourist attraction in Christchurch. This is the closest most people ever get to experiencing Antarctica, and ironically, it turned out to be the only place during our trip in which we saw penguins. The Little Blue Penguins (*Eudyptula minor*) housed in the facility were nowhere to be seen on the ice shelf close to McMurdo. In fact, their range is limited to Australia, where they are sometimes called "Fairy Penguins." Antarctic penguins include the giant Emperors, Adelie, southern Rockhopper, King, Chinstrap and Gentoo, but none were to be seen during our stay. It turned out that the ice shelf was very extensive in the 2016-17 season, too far for penguins to travel. Worse, any close enough to be seen near McMurdo would probably never find their way back to sea and would consequently perish. Thus, we were not disappointed to miss out on such a sighting.

The International Antarctic Centre also contains a replica of Scott's cabin in its "4 Seasons of Antarctica and Explorer Legacies" exhibit, where visitors learn about the stories and hut legacies left by Scott and Shackleton and experience a light and sound show of the four seasons of Antarctica. Among other experiences is a Hagglunds ride, which mimics the amphibious vehicle used to cross ice shelves. We used one during our stay to go out onto the ice shelf near McMurdo when we did our crevasse rescue training. Another popular attraction is a room in which you can experience a simulated Antarctic storm (gear is provided). Rosaly wanted to try it, but Mike figured they would be seeing enough of the real thing.

Fig. 3.4. The Little Blue Penguins at Christchurch's International Arctic Centre are a main attraction (Photo by the authors)

Flying South

One must always be prepared for the unexpected when traveling to Antarctica. We learned that two of the scheduled passengers who had been in the same group as us picking up gear the day before would be turning around and heading back home. The project they were on had to be scaled back at the last minute due to unforeseen constraints, and so they were no longer needed. We could not imagine their disappointment.

Our flight south departed Christchurch on the scheduled date (called the "Ice Date"), which seasoned Antarctica-bound travelers say was lucky. It is common to have to wait a day or two and sometimes longer. We wondered if our luck would hold or whether we would be boomeranged. We kept track of time during the flight and knew when we had reached the point of no return. As it would turn out, we were exceptionally lucky with weather during our whole trip. There were a few gray bad weather days, but they hardly affected our program. One was perfectly timed to happen on a day when we were acclimating on Erebus, during which we were staying put and not doing much activity anyway!

The departure from Christchurch felt like a grand adventure. The Antarctica terminal is draped with a multitude of international flags, and just to be there is thrilling. Everyone has to wear full cold weather gear on flight, just in case of an emergency landing on the ice.

We and all our luggage were weighed. After waiting while the "checked" luggage was stowed, we boarded a bus and got to our 4-propeller LC-130 Hercules aircraft, the work-horse of Antarctica. The ice-bound aircraft, specially equipped with skis, gets the added "L" in its designation. Some aircraft in the flight group are in the process of getting new engines called Scimitars. Eight curving blades replace the four straight propellers of the older design, making them more efficient for ski landings and more powerful for takeoffs without JATOs.

Inside, we took seats along the walls of the plane in typical troop-carrier fashion. We noticed 5-gallon water jugs hanging from a central structure that housed a row of unused seats. The National Guardsman on duty (acting as "flight attendant") passed out generous lunches: two sandwiches, a candy bar, granola bar, cookie, 2 bags of chips, and a fresh apple. The cargo master piled and secured our carry-ons down the center of the cabin. Toward the back of the plane, just where our cabin-length canvas benches ended, pallets towered up to within inches of the ceiling, where small round blue lights illuminated their surroundings. Also back among the cargo was our bathroom facility, a curtained toilet seat over a bucket. We took our seats and rested our backs against red netting. With legroom to spare, the flight was more comfortable than many of today's commercial airlines. The crew warned us to not leave our luggage on the floor if it contained anything that would freeze. Most backpacks hung from the walls on fasteners. Our hosts also passed out ear-plugs. There would be no conventional conversation on this noisy flight; we communi-cated by typing on our laptops.

Fig. 3.5. Exterior and interior views of our LC-130. The cargo hold afforded more legroom than most commercial flights do (Photo by the authors)

Once airborne, we felt as though we were in the cast of an old WWII movie about the courageous crews in a B-17, braving high-altitude cold in what is basically a flying warehouse. The ceiling and all the walls were festooned with cables, copper tubing, and metal shelves holding strange equipment with fat red dials and silver switches. First aid kits swung back and forth in a row above the windows. Mirroring them just below dangled the life vests. Behind our seats, oxygen hoods nestled in little bags in case of fire or decompression, which can happen even at the relatively low altitudes followed by the Hercules.

Our seats were not directly in front of the small portholes, so we saw nothing until we stood up and peered through the ones accessible to us. We were free to move about the cabin, and the views of the land, sea and ice from the portholes were incredible. Soon, the green coastline of New Zealand gave way to unrelenting clouds just below the aircraft. Through occasional breaks we could see the steely sea undulating in great, sluggish waves. Heat blew from two big pipes on the ceiling. Five hours into the flight, it began to get chilly. One passenger near us quipped, "You think that's cold? I guess we ain't seen nothing yet."

Fig 3.6. Three views out of a porthole in the LC-130. Thin sea ice floated in shapes that reminded us of water lilies (Photos by the authors)

Some travelers were initially disappointed that they were taking the LC-130 Hercules rather than a jet (which can only be used early in the season when the ice runway is hard enough for wheels). Flight times for the jets are about 5 hours; the Hercules would take closer to 8. But slower aircraft fly low and slow over the gleaming Antarctic landscape upon approach, a payoff for patience during the long flight. The flight treated passengers to 45 minutes of glorious views out the tiny portholes: glaciers, icebergs floating in "water lilies" of sea ice, and rugged black mountains rearing out of the brilliant white ice all paraded by. Often, we spent so much time peering at the brilliant white views outside that when returning to our seats, our eyes took some time to adjust. It is a good idea to wear sunglasses when gazing out the portholes.

Because we could not see out of the portholes while strapped down, the landing was a bit mysterious. We had to guess what was going on by listening. We could also see up against the ceiling twisting tubes leading to the flaps as the pilots raised and lowered them. The landing was gentle, but the scrape of the snow and ice on the landing skis filled the cabin with a rumbling hiss. We stepped out to find Erebus towering behind the aircraft.

4

Second Step: McMurdo

Fig. 4.1. *Mars Climb* by Michael Carroll

M. Carroll, R. Lopes, *Antarctica: Earth's Own Ice World*, Springer Praxis Books,
https://doi.org/10.1007/978-3-319-74624-1_4

The final moments of a flight to McMurdo aboard a Hercules are at once exciting and frustrating. After seeing the rugged mountains and blinding flat plain of the Ross Ice Shelf from the air, we know what awaits us just outside those netted metal walls. Those of us who have never been to the White Continent don our full set of our survival gear, from boots to balaclavas. The seasoned travelers look sideways at us, a gleam in their eyes. Some bundle up like us, while others wear pre-approved casual winter jackets, unzipped.

The doors crank open and light floods the interior. A blast of cold air tsunamis over us – not bad for a Colorado boy, but enough to make the Ipanema girl shiver. We step out into a glistening world of sunshine and ice. The Hercules rests in the middle of an open white plain, behind which rises our ultimate destination: Mount Erebus. The mountain is visually stunning, rearing up over 3,700 m from our vantage point at sea level.

Fig. 4.2. Rosaly Lopes in front of our LC-130, which rests safely on the Ross Ice Shelf. Mount Erebus is the mountain behind (Photo by the authors)

The crew is already hard at work offloading McMurdo's supplies onto tractors. Guides usher us to a huge red bus emblazoned with the words "Ivan The Terrabus," which we are later told is a local favorite. The ride to McMurdo takes us across the smooth Ross Ice Shelf to the coastline of Ross Island, a gentle climb from frozen ocean onto frozen land. The rocks and dirt are dark brown, ground volcanic rock. Once on terra firma, we follow a parade of flags marking the route to civilization. Red and green flags delineate directional routes. Should the weather go south (and it doesn't have far to go), these flags are our only hope of a safe return in whiteout conditions. Blue flags mark fuel lines strung across the surface of the ice and snow. Black flags warn of danger: hidden crevasses, thin sea ice, or restricted wildlife areas.

We pass the stunning green buildings of Scott Base, New Zealand's stellar center of operations. Our bus stops to drop off some people at Scott and then continues two miles farther down the coast, where the road ascends between two volcanic shoulders. We drop down into the storied McMurdo Station, one of the most remote outposts in the world and the largest base in Antarctica, perched upon Ross Island's McMurdo Sound. The community of McMurdo Station faces roughly in the direction of New Zealand, one of the nearest inhabited spots of land (only Argentina is closer to the continent, but on the opposite side). From a winter population of less than 200, the town's ranks swell to nearly 1,000 at the peak of the austral summer. McMurdo has its own general store, hair salon, gym, coffee house, bar (*Gallagher's Pub*), chapel, public library, and an extensive galley, open 24/7 to serve people in all work shifts. The typical work week in Antarctica is 54 hours, Monday through Saturday.

The galley offers a variety of options, including fresh-baked cookies, pizza bar, grill, deli and hamburger joint. Food at McMurdo is free to all USAP workers, scientists and grantees (if you are a US citizen, your taxes have already paid for it). Meals are served cafeteria style. During the summer season, diners are asked to give priority to night workers during the busiest time (roughly 11:30 pm to 1:00 am). Food service management can make arrangements for takeout meals for those working in the field or those who are ill.

The only ATMs on the continent reside at McMurdo. There are two machines (only one of them usually works). Any visitors to the South Pole Amundsen/Scott Station must bring enough cash to last the duration of their stay, and Palmer Station is cashless. Credit cards cannot be used at South Pole Station.

While beginning as a US naval air facility (see Chapter 1), McMurdo now operates under the administration of the National Science Foundation's United States Antarctic Program (the USAP). Yet, as is true of all outposts on the southern continent, McMurdo is not sitting on United States real estate: No one owns any part of Antarctica.

Roughly 3,000 participants work at US Antarctic facilities. Of these, 90% travel through New Zealand to reach McMurdo. The rest sail from Punta Arenas, Chile, to Palmer Station on the opposite side of the continent. USAP participants come from nearly every state in the US, with the majority coming from Alaska, California, and Colorado. 80% carry out their work during the austral summer season (roughly late October to early February). Some 700 scientists perform investigations on more than 100 different science projects.

McMurdo is a village braced against the elements. Every building has doors equipped with latching handles for a tight seal. A blazing sign outside of the galley and general store displays constantly updated weather conditions. A large monitor near the main doors tracks flight information to and from various field camps and the mainland. Often, the schedule board is filled with flights in red letters, indicating cancellations. Any cancelled flight triggers a domino effect involving other teams and their transport. Travel to the South Pole can be especially frustrating; delays of days or even weeks are common. Often, an aircraft is fully loaded and waiting when its travel is cancelled due to weather, scheduling, or mechanical issues. But seasoned travelers in the south keep a philosophical attitude. Teams planning to go to remote sites understand that patience is needed and, for the most part, take a humorous view of unfortunate situations and try to make the best of their time while waiting at McMurdo. Typical of this outlook is a sign displayed on the notice board in one of McMurdo' office buildings from a remote site called Yesterday Camp. (The camp gets its name from its proximity to the International Date Line: just walk a little to the left and you are in the previous day.) The sign reads:

"Antarctica's premier destination for tears, hopelessness and no LC-130s. There is no tomorrow at Yesterday Camp."

Mac Town, as the inhabitants affectionately call it, has the feel of a small town. People watch out for each other, care for the common safety and health, preserve the delicate polar environment, and even make sure everyone has an occasional dose of entertainment through local bands and parties (most notably, New Year's Eve's "Icestock"). Guidelines for new arrivals describe anticipated duties: "At times, everyone may be expected to work more hours, assist others in the performance of their duties, and/or assume community-related job responsibilities." Those community-related responsibilities include everything from scrubbing floors to disinfecting common areas in the dorms, cleaning bathroom areas or washing dishes.

In addition to visiting scientists, the fascinating personnel at Mac Town include experts in communications, engineering, mechanics, medicine, power generation, waste management, mountaineering, environmental safety and protection, and specialized field staff.

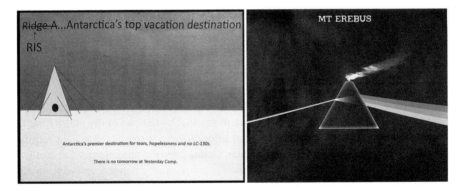

Fig. 4.3. Two typical posters found in the corridors of Mac Town (Photos by authors)

Fig. 4.4. Panorama of McMurdo Station. At left, the NSF's Elaine Hood points out landmarks. (Photo montage by Michael Carroll)

McMurdo has its own medical facilities, which are impressive considering the remoteness of the station. Germs are of extreme concern. McMurdo is a closed environment with signs constantly reminding people of hygiene and correct disposal or recycle of refuse. A hand-washing station greets all visitors on their way to the galley, serving as a social meeting place as well as a cleansing center. As Heather Dodds, McMurdo Hospital's lead physician for the 2016/17 season, puts it, "If something bad came up like the influenza, then hygiene really matters, whether you are in Antarctica or Connecticut. The difference is that we don't have the resources to deal with a pandemic [at McMurdo]. Think of people on subways in New York City: they're in close contact and we don't have discussions about it. But in McMurdo, things are very controlled. We have detailed standard operating procedures for everything, and it really works; it absolutely cuts down on contagions to wash your hands, cough into your elbow, etc."

Sickness can endanger operations on many levels, as NASA learned with the Earth-orbiting flight of Apollo 7. Apollo 7 was the first piloted flight of the craft that would take humans to the Moon. Just hours after reaching orbit, Commander Wally Schirra came down with what appeared to be a common cold. But in microgravity where sinuses don't drain well, there is nothing common about a cold. The only way to clear sinuses is to blow the nose, and this puts added strain on already congested eardrums. Within 24 hours, illness also visited senior pilot/spacecraft navigator Don Eisele.[1] Understandably, the combination of a confined capsule and the distress of illness caused some stressful moments of communication between mission control and the crew. In distant settings such as an outpost on Mars or the Moon, precautions must be taken for even the common cold, among more serious ailments.

McMurdo General Hospital handles more than the common cold. Antarctica presents a rough and sometimes treacherous environment from natural forces to industrial dangers, but the NSF has put into place a host of excellent guidelines and safeguards for operations. "When I first got [to McMurdo], I was a bit skeptical of the need to emphasize safety to the extent they did," says Dodds. "But I have to say that by the end of my entire time there I was impressed with how few work-related injuries there were. The constant reminders effectively reduce the incidents." As evidence, Dodds cites the annual vessel resupply that occurs at the end of Antarctic summer, which is a potentially dangerous event involving heavy machinery and the delivery of equipment and supply crates. "In the entire vessel offload, there were no injuries. Safety was always the focus."

Visiting researchers are provided office space and laboratory stations in the Crary Laboratory, which is designated "Building 1." Built in 1991, the impressive facility is named after Albert P. Crary, a respected researcher who studied geophysics and glaciology. Crary was the first person to step onto both the North and South Poles in 1952 and 1961, respectively. The Crary Lab is the largest science facility on the continent, covering 4,320 sq-meters on three levels. Laboratories include areas for marine and microbial biology, earth sciences like geology and volcanology, and atmospheric and space sciences. The laboratory's technology is state of the art, rivaling any major research university lab.

---

[1] Walt Cunningham dodged the germy bullet, although official reports said that all three astronauts came down with the bug. Cunningham quipped that "on Wally's crew, if the commander had a cold everyone had a cold."

Crary has a volcano observatory center, which monitors seismic activity at Mt. Erebus. Also included in the lab's layout is a darkroom, freezers for preserving and analyzing ice cores, and a sophisticated electronics atelier. A library serves as a lecture hall, which hosts weekly lectures from McMurdo's residents and visiting scientists. Display cases feature fossils and meteorites from the southern continent. Three extensive aquariums, including "Touch Tank" (see Chapter 7), exhibit Antarctic marine life from the ocean just outside.

The laboratory perches on a slope that leads down toward the McMurdo Sound below. Crary's main floor, also called Phase I, is primarily used by biologists and chemists. It houses microscope benches, environmental chambers, and freezer rooms. This level also contains cargo areas that warehouse equipment storage and carpentry workbenches. The building stairsteps lead one to the next level down, Phase II, which contains workspaces for geologists and glaciologists. The lowest level, Phase III, includes the aquariums and the marine sciences research areas.

Playing Nice

Antarctica is a model of international cooperation. Personnel are frequently exchanged among the various stations scattered across the continent. Many large science projects are coordinated, sharing logistics and resources among international partners. Exchange or shared use of ships and aircraft is a common occurrence. COMNAP (Council of Managers of National Antarctic Programs) links resources between the 29 signatories of the Antarctic Treaty. At McMurdo itself, roughly 30% of the power is wind-generated, but this power is split with nearby Scott Base, the New Zealand station. New Zealand also shares the heliport at McMurdo and coordinates research and logistics. In the southern wastelands, cooperation is a matter of survival. Science stations spread out across the frigid wilderness from countries as diverse as Japan, Finland, Argentina, Russia, Brazil, and Norway.

Antarctica's location makes the continent ideal for research in fields including astronomy, atmospheric and magnetospheric sciences, paleontology, marine and microbiology, and even meteorite searches. The Scientific Committee on Antarctic Research (SCAR) coordinates international research across the continent. The group is independent of any specific government or nation and serves with a "view to framing scientific programs of circumpolar scope and significance." SCAR organizes symposia, publishes annual reports, develops long-term plans, and carries out other activities designed to foster the regular exchange of information and news pertaining to scientific programs and research. Antarctica is both a perfect model for terrestrial cooperation and a prototype for what may be needed at future Martian or lunar settlements. And herein lies perhaps the most dramatic planetary analog of all.

To Erebus, by Way of Mars

For a preview of a Mars settlement, the US Antarctic Program's McMurdo Station is hard to beat. Anyone who has studied architectures for Martian colonies will recognize common elements here. McMurdo is a remote, sequestered setting, essentially cut off from the infrastructure of the rest of the world. It must be able to function independently for long periods of time, generate its own power, and handle its own emergencies while at the same time enabling researchers to carry out work in its bleak surroundings. After visitors arrive at McMurdo, one of the most striking Mars settlement analogs that they notice is the arrangement of the doors: they are all configured like airlocks with outer and inner accesses. Inner doors are not open until the outer ones are sealed. Door handles are the horizontal type used on freezer doors, because winds can reach hurricane-force and entries must seal completely. The difference with these freezer doors, of course, is that the cold is being kept out, not in. On Mars, an airlock will provide a transition from the thin air of Mars to the comparatively high pressure inside human habitats.

Fig. 4.5. (*Left*) A blue, double-doored "air lock" at the front of a rigid tent near the Lower Erebus Hut on Mt. Erebus; (*Center*) A typical "freezer-style" door handle in Antarctica; (*Right*) An astronaut exiting the Quest airlock on the International Space Station (Left, center photos by the authors; right photo courtesy NASA/JSC)

Doors are just the beginning of McMurdo's strong parallels with Mars' base designs. The reason is simple: form follows function, and McMurdo functions in a very Mars-like environment. Surrounded by hostile terrain, glacial floes, and some exotic, deadly environs, Mac Town is a settlement where people recognize the hostile nature of their location at all times.

Antarctica is the coldest, windiest, driest, deadliest place in the world. An ice sheet up to three miles deep covers the continent, a landmass larger than the continental United States. This ice constitutes roughly 90% of the world's fresh water. Marine life along the coastlines is varied and abundant, ranging from seals and penguins to Skua birds and Killer Whales. But the continent's barren interior is nearly sterile. "You can be out in the middle of [an ice sheet] and find not a microbe in sight," says NASA Ames astrobiologist Chris McKay, speaking of similar terrain in Greenland. "You don't see life until you get to the edges where there's liquid water."

The Martian surface is even more sterile than Antarctica's: hydrogen peroxide laces its dirt and the unchecked solar wind continually bathes the landscape in radiation. On Earth, surface radiation is mitigated by our planet's magnetosphere, a magnetic "bubble" that shunts most of the Sun's radiation away from the surface. At the poles, some of this energy funnels down along the planet's magnetic field lines. This confluence of cosmic energy interacts with Earth's upper atmosphere, causing the famous Aurora Borealis in the north and the Aurora Australis in the south. These curtains of beautiful light pay tribute to the fact that the Earth's surface enjoys protection from outside radiation thanks to the energy fields generated by its own molten iron core. Whereas the Earth's core acts as a gigantic bar magnet producing lines of energy around the planet, the Martian core is smaller than the Earth's and has probably cooled to the point that very little magnetism is generated today. The only magnetic fields, left over from earlier epochs when the heart of Mars was warm, come from localized rock in the crust. The Martian magnetic field is 1/40 as strong as Earth's, far too weak to protect the surface from infalling radiation.

Like Mars, Antarctica is a desert. Although covered in frozen water, scant precipitation falls annually. Averaged over its entire surface, a measly 166 millimeters (6.5 inches) of moisture falls each year, a low enough draught level to qualify the continent as a classic desert. Conditions are even more arid in its dry valleys, where precipitation seldom exceeds 100 mm (4 inches). Antarctica's coastlines tend to get more. Precipitation is almost exclusively snow, with the exception of some coastal areas where rain falls at the height of summer.

The Harsh Continent is a tropical paradise compared to Mars, where the only precipitation today falls as dry ice (frozen carbon dioxide) over the polar regions. The Martian air has enough water vapor to form clouds in some areas under transient conditions, while frost forms on the ground in the higher latitudes. Antarctica's chilled skies often display haloes, sundogs, and other phenomena related to ice crystals suspended in the air. These beautiful spectral sights may be a common occurrence on Mars as well, since dust and ices are suspended high in the Martian atmosphere.[2]

Although Mars' rarified air is only as dense as the Earth's at 100 km above sea level, Martian air currents do build up winds, another aspect that the red planet shares with Antarctica. With its low air pressure, no Martian winds will be blowing any doors open. (Andy Weir, author of the popular book and film *The Martian*, was well aware of this but used a violent windstorm as a dramatic element in his story). However, the difference in air pressure between the Martian environment and any inhabited human areas will require care in traversing those double-doored airlocks.

Travel even within the confines of McMurdo Station has other unique requirements. Like astronauts, inhabitants of McMurdo must "suit up" before going outside. All NSF personnel, grantees, and researchers are expected to don a minimum set of specific clothing before venturing out, even if the weather appears to be fair. The capricious meteorology of the southern continent can endanger those who are not prepared for it. The standard-issue ensemble is called the Extreme Cold Weather (ECW) gear. The protective set of clothing includes special boots, hat, goggles, balaclava, gloves, Carhart bib, insulated pants, and most visible of all, the "Big Red" coat (a lighter version is known as "Little Red" and, yes, the red coats have a hood). Layers of clothing for the upper and lower body are also supplied as options.

According to the NSF official guide, "The ECW clothing is functional, sturdy and cost effective. It includes special items of outer clothing required for the Antarctic climate,

---

[2]For more on Martian weather, see *Drifting on Alien Winds: Exploring the Skies and Weather of Other Worlds* by Michael Carroll (Springer 2011)

Fig. 4.6. An ice crystal halo crowns the Sun over Fang Glacier, an ice field at 2,745-m altitude (Painting by Michael Carroll; collection of Rosaly Lopes)

such as parkas and boots. The majority of clothing is in men's sizes, but will fit both men and women...the type and amount of clothing you receive depends on where you work and what your job title entails. Most, but not all, of the ECW clothing is mandatory." Astronaut Nicole Stott, flight engineer on the International Space Station's Expedition 20 and 21, believes the parallel between Antarctic gear and EVA spacesuits is a good one. "I think the

Fig. 4.7. (*Left*) Nicole Stott conducts a spacewalk during the STS-128 mission (NASA/JSC); (*Right*) Two explorers in their ECW gear. Suiting up on Mars would feel a lot like a foray at the ISS or McMurdo (photo courtesy Evan Miller)

analog is totally reasonable. It's a lot like how we use the undersea / Aquarius missions as an analog to spaceflight. As part of those missions, any time we go out on SCUBA or the helmet/umbilical dives, we go through a protocol very similar to an EVA—in fact, we treat them just like EVA's with the same suit-up communication between crew members and our ground (topside) control teams." During her 6 ½ hour spacewalk, Stott wore some 254 pounds of suit and equipment, enabling her to carry out the construction of ISS elements as well as the retrieval of science experiments. One can envision a future base on Mars, where in addition to protective clothing that offers warmth, Mars inhabitants will also require oxygen and a pressure or support (skintight) suit. In Antarctica, the air temperature reminds prospective explorers of their sartorial needs. On Mars, the deadly low pressure will serve the same role.

The winds on Mars will generate something very familiar to inhabitants of McMurdo: dust. Dust can be a serious issue with mechanical devices and delicate scientific equipment. Lunar dust posed problems to the Apollo astronauts, who at times had difficulty resealing interfaces between gloves and sleeves, suit collars and helmets, and operation of various tools like cameras or drills. Lunar dust is essentially powdered Moon dirt. The dirt on places like the Moon and Mars is referred to as regolith, meaning ground up rock that lacks the organic material found in terrestrial soil. Antarctica's dirt is very much like regolith, with almost no organic material interspersed with the rock, gravel, and sand of the ground. Dust at McMurdo can be fine. Summer dust storms blow through the streets of Mac Town, powdering surfaces and structures with a dark brown blanket. Aside from fouling equipment, there are other problems inherent in the dust on the Moon and Mars. Moon dust has been found to be fairly toxic. It is angular and abrasive, posing inhalation hazards. Martian dust is highly corrosive. Future colonists will need to take special care to ensure that dust does not get tracked into their living environment.

Dust in a terrestrial setting can also lead to a host of medical problems associated with eye infection, asthma, and other breathing complications. Dust was initially of prime concern to those traveling to the continent with asthma, but in recent years, the heavy equipment operators at McMurdo have begun to use more water on the roads in town to reduce dust. In Antarctica, one change has reduced incidents of this type to nearly zero. "Those with a history of uncontrolled asthma/COPD were not PQ'ed [physically qualified] to come," Dodds comments. "Otherwise, I am sure those with more progressed underlying lung pathology would have been showing up at the clinic with 'exacerbations'."

The USAP guidelines state:

> *Antarctica is an extreme, remote environment, and medical facilities are limited. U.S. Antarctic Program facilities are equipped and staffed to provide routine ambulatory care that would be expected in a U.S. clinic, and have the capability to stabilize and manage a range of emergency medical and dental conditions before transporting patients off the continent. However, medical evacuations take a lot of time and effort and place others at risk, even when the weather allows travel. Remote field camps and research vessels pose additional difficulties.*

In the most critical cases, personnel must be medically evacuated to the mainland (usually Christchurch), but medevacs take time and are costly. Incidents in remote field camps or on research vessels pose further complexities. The PQ process is designed to screen out those whose medical conditions cannot be effectively managed on the ice or aboard research ships. "When I worked in Uganda," Heather Dodds explains, "you're dealing with life and death every moment, so you just do your job to heal the best you can." Sometimes, standard medical practice might entail prescribing a medication that may take time to work, or waiting to see if a patient's issue will resolve on its own. In this sense, Dodds suggests, "Medicine is the same whether you're in Portland or Alaska." Yet McMurdo Station presents a unique set of circumstances. It is distant enough from civilization that medical issues can become emergencies far from infrastructure, but close enough to the mainland (i.e. facilities in Christchurch or Sydney) that a patient may be shipped out for care. The tension creates pressure on medical experts, both overseers at the UTMB and NSF, and the medical personnel on site at McMurdo. If decision makers approach a given situation conservatively, they may decide to have a person evacuated to the mainland rather than taking the wait-and-see approach more common to facilities in a population center away from Antarctica. It's an expensive but sometimes prudent prospect. But once medical personnel make landfall on distant planets and moons, they'll experience a major difference – for better or for worse – between their remote hamlets and McMurdo's community: that of total independence. "Mars will work better in some ways," Dodds argues, "because you won't have the option of transferring people to a facility on the mainland. What you have on hand is the only option you have."

Long-Distance Calling

Within McMurdo Station, a telephone system serves the entire community. Cable television comes via the American Forces Antarctic Network, which offers six channels. Argentina, Chile and the US all have local radio station broadcasts. Internet is also available, although individual access is restricted at McMurdo to preserve the limited bandwidth. At the time of our visit, the telecommunications are serviced by Lockheed Martin, based in Denver, Colorado. This means that a phone call to Colorado some 14,000 km away is considered a local call.

A communications hub for McMurdo is located on Black Island, though the plan is to move all communication equipment to Ross Island over the next several years. The primary reason for a station on Black Island is that Erebus blocks the view from most places on Ross Island. At latitudes that far south, the dishes need to be pointed almost horizontally, and it is difficult to find places on Ross Island not blocked by the mountain. However, a station on Black Island presents its own set of problems. The station only houses personnel during the summer months and runs unmanned for the rest of the year. If equipment problems require staff to go there during the winter, they have to travel some 40 km across the ice in darkness and possible bad weather to reach it.

As future Martian bases and outposts grow into settlements, telecommunications systems will naturally grow as well. Farther afield, any journey into the Martian wilderness will need to be as carefully supervised as it is in Antarctica. At McMurdo, all travel, whether by foot, snowmobile, tractor, or aircraft, must be registered with the command center, "Mac Ops," where radio communication is constantly monitored. Communications personnel keep tabs on all those traveling in vehicles, including fixed-wing aircraft, helicopters, and local and long-range ground transport. When scientists or others embark on sorties of any kind, they must confirm plans with Mac Ops, detailing their projected time of departure, beginning of return trip, and expected arrival time. If the trip is to take more than a day, a daily check-in time is established at the outset. At a remote location, a "put-in" call is required before the delivering vehicle or aircraft leaves. This call includes the location name, the name of the camp leader, the number of people (which is relayed by the event number of the project), and confirmation of the daily check-in time. Each daily check-in requires the location name, the number of people involved (by the National Science Foundation's assigned event number) and the "all is well" message.

The stakes are high on this schedule. Should a team be late for check-in, a chain of events automatically triggers. After one hour of waiting, the Emergency Operations Center (EOC) is activated. The EOC mobilizes a host of people, including the NSF Station Manager, the Emergency Communications Manager (who alerts a network of people to monitor radio transmissions), the Science Support Manager, Information Technology Manager, and the Fire Chief. These people are in charge of making a plan for rescue. When it comes to check-in, promptness is preferable to an expensive and embarrassing sequence like this.

McMurdo relies on several modes of long-distance communication. Iridium satellite phones are in operation at the pole and most remote field camps, and can be used at sites like Fang Glacier and Lower Erebus Hut.

Power, Water, and Waste

The notorious winds that can make it painful to walk between buildings at McMurdo are useful for the station's power needs. Engineers have erected a parade of wind turbines on a ridge that rises between the McMurdo and Scott bases. These turbines generate over 30% of the power needed at both stations. When winds are brisk (which is often), the wind turbines and one of the smaller generators could supply all of McMurdo's power needs. However, Antarctica's winds are capricious and erratic, so other sources of energy must be added to the mix. Four large Caterpillar diesel generators supply the majority of McMurdo's electricity, with two smaller ones as backup in a separate building. When power is interrupted or winds are calm, the big generators kick on and generate what's needed. The station experiences peak loads during the day when the most work takes place (this is true even in summer, when the sun never sets), with the most severe load requiring upwards of two megawatts of electricity.

Diesel power will not be an option on Mars. Diesel fuel requires oxygen to burn, and in a non-oxygen environment like Mars, there are better options. McMurdo's wind turbines would not be nudged in the fiercest of Martian dust storms, but ultralight wind turbines using super strength, advanced materials to construct huge blades may be an option. Solar panels dot the landscape around various remote camps in Antarctica, where portability trumps the variable light levels of daylight hours. Solar power is a source that has been used successfully by many Mars vehicles, and nuclear power has an even higher and more consistent yield, as it can be run 24/7. In addition to day/night cycles, nuclear power is also unaffected by Martian dust, which can significantly reduce solar panel output. The Curiosity Rover is nuclear powered and has done exceedingly well in the Martian environment.

Fig. 4.8. Two sources of power in Antarctica: (*Left*) solar panels situated at the Lower Erebus Hut and (*Right*) a row of wind turbines that supplement power for both McMurdo and Scott Bases (Photos by the authors)

Water for McMurdo's inhabitants comes directly from the sea. A sophisticated water desalinization plant purifies seawater through a reverse osmosis process. The water in all faucets, drinking fountains, and showers is potable. While in McMurdo water conservation is encouraged in many ways. With up to 1,000 residents and visitors at the peak of the season, wastewater is still generated in prodigious amounts. The advanced design of McMurdo's water treatment plant makes use of anaerobic bacteria to break down sewage in the beginning of the process. From there, the wastewater flows through bins containing aerobic bacteria that further break down the sewage. By the time the fluid makes it to the far end of the plant, it is nearly clear. At that point, it is filtered and sterilized with ultraviolet light. The water, now essentially pure, is sent back into the ocean. Solids collected by the filters are pressed into bricks and dumped into crates, where they await shipment to the United States for disposal.

According to the USAP official literature, "What comes in must eventually go out." To that end, McMurdo is one of the most efficient recycling communities in the world. Nearly 100% of its garbage is shipped off the continent for appropriate disposal back in the US. Inhabitants sort trash into a dozen different categories (i.e. glass, food waste, mixed paper, plastics, cardboard, aluminum, clothing and rags, paper towels, hazardous material/non-recyclables, etc). The various materials are forklifted to the Waste Barn where they are sorted again and compacted. Refuse bound for the US travels in 20-foot long milvan (Military Owned Demountable Container) shipping containers, each containing up to 40,000 pounds of material. McMurdo's trash collection originates from farther afield than *Gallaghers Pub*: the station processes refuse from all the field camps and the Amundsen-Scott South Pole Station (although some human refuse at the South Pole is buried). The refuse is stored in giant crates in an industrial-looking section on the hills above the center of town. There, it awaits its annual shipment to processing in California at the end of the season. For McMurdo garbage, everything goes.

Recycling of water at a Mars settlement will be critical. Getting added water from the environment is tough. The limited Martian water resources will essentially be mined, with solid ice dug from the subsurface. Permafrost is notoriously dense and hard on equipment, so advances in drilling and mining operations will need to be made in order to quench the thirsts of such a remote community. Water vapor can also be culled from the atmosphere as gases are processed for other uses, but it is not abundant. As in McMurdo, conservation will be the order of the day. Antarctica's solution to garbage disposal—shipping it home—will not be feasible on this distant world. Studies have been carried out to find a solution, with some suggesting that inhabitants create a Martian landfill and compact trash into bricks for use in construction.

Staging to Field Camps

McMurdo serves as staging area for many remote outposts and field camps. From the glacial valleys of the Transantarctic Mountains to the ice plateau surrounding the South Pole, field camps and science stations dot the continent's landscape. Caches and fuel depots stand in strategic locations, serving as lifelines between the outposts and the relative civilization of McMurdo, Scott, and other permanent bases. An entire continent's infrastructure depends on this system. The point was dramatically demonstrated in the 2016/17 season, when weather prevented flights into the wilderness early in the season. The ripple effect of shortages was felt throughout the season as expeditions had to be scaled back, changed to other locations, or cancelled outright.

Like McMurdo and other large year-round facilities, Mars settlements will serve as staging areas for extended exploration of the Martian frontier. From a central, safe location, future explorers will depart for the poles, the great canyons, and the soaring volcanoes of the red planet.

McMurdo serves as the home base and departure point for an assortment of field camps (see Chapter 5). These operations are facilitated by the Antarctic Support Contract (ASC), a business division of Leidos. While the National Science Foundation manages the US Antarctic Program, a variety of companies provide logistical support. ASC hires people to cook food, repair diesel engines, and supply fuel for aircraft, all the while ensuring the success of research projects. Once the National Science Foundation has provided a grant to a researcher, ASC planners work closely with the NSF to determine the feasibility of the proposals. They establish what resources will be needed, such as helicopters and snowmobiles. They check time frames, confirm where researchers need to go on all portions of their trip, and establish whether the necessary logistics will be in place at the right time. Once the planners work out these big-picture schedules, other ASC personnel known as "implementers" then take responsibility for ensuring the project's success "on the ice," coordinating travel into the hostile Antarctic environment by scheduling flights and surface trips from McMurdo. According to implementer Elaine Hood, "We do the more detailed planning to ensure that you can get done what you need to. Our job is to get you where you need to go and to ensure that you have the supplies you need once you're in Antarctica." While implementers carefully assist the groups that deploy into the field, many science teams go multiple times or have team members who have been there before, so they have developed a built-in support system. These adventurers know their way around. When they land in McMurdo, they know where to find their dorm room and the cafeteria and how to schedule their flight to the field camps. In the case of the Artists and Writers program, most have never been to the southern continent. As Elaine Hood told us, "Artists and Writers get more attention from the implementers because they have never been here before and will probably never return. They almost always are on their own with no teammates. They have a very short window of opportunity to accomplish their project, usually just a few weeks. It is most efficient for everyone if we provide close guidance and support."

Antarctica throws many challenges and roadblocks in the way of the implementers, from mechanical failures to weather crises. As Hood puts it, "we have problems and issues that we constantly have to resolve. Nothing is routine and boring in this job." Through the efforts of implementers like Elaine Hood, hundreds of people are able to carry out research annually in the most remote corners of Antarctica's frozen wilderness, from glacial meteorite fields to the Mars-like Dry Valleys. Tomorrow's explorers will need to rely on such people as they establish base camps and far-flung outposts on the Moon, Mars and, eventually, the moons of the outer Solar System. The people who support, train and encourage Antarctic researchers will have their counterparts at the beachheads of Syrtis Major, Elysium, Valles Marineris, and beyond. We will need them.

Homes Away From Home: Imagining a Mars Outpost

The harsh Martian environment will offer many challenges to designers intent on setting down permanent outposts or settlements. In addition to corporate analyses done by major aerospace firms, two private groups in particular have conducted detailed Mars colony studies. Florida's Four Frontiers Corporation has carried out several design studies, resulting in multiple scenarios for Mars settlements. The Mars Society, with roots in the famous *Case for Mars* conferences, has executed complex design studies detailing many aspects of Mars outposts, including greenhouse technologies, habitats, power generation, and even entry vehicles. The Society has actually erected Mars simulation habitats in the Utah desert and in the Canadian arctic. Mars Desert Research Station (MDRS) is located in southern Utah, while the Flashline Mars Arctic Research Station (FMARS) stands on Devon Island in northern Canada.

Fig. 4.9. One Mars Society study envisions a biconic entry vehicle descending on parachute towards an advanced settlement. Habitats are buried beneath a meter of dirt for radiation protection. Vast greenhouse assemblies generate food and replenish oxygen for the inhabitants. In a nearby crater, we see the twin cooling towers of a nuclear reactor, which produces energy for the settlement's activities. A communications center crowns a butte at right. (Painting by Michael Carroll)

In 1984, attendees at the Case for Mars Conference held in Boulder, Colorado outlined baseline approximations of a Mars settlement and the infrastructure needed to service such a distant community. Aerospace engineers, planetary scientists, journalists, and artists attended the meeting, contributing a diverse trove of resources and input to the project. The resulting scenarios continue to grow today under the auspices of the Mars Society.

Key to the Society's baseline scenario is a strategy called "Mars Direct," the brainchild of aerospace engineer and Society president Robert Zubrin. Rather than have massive, expensive launchers with huge human-tended vehicles, Zubrin's approach envisions a series of sorties that build up infrastructure over time. As fuel is one of the largest limiting factors in any Mars scenario, Zubrin proposes a vanguard of unpiloted craft arriving at Mars with empty fuel tanks on their built-in Earth return vehicles. These tanks would gradually fill with fuel manufactured on the spot, using *in situ* resources from the Martian air. When the second and third ships arrive two years later – one carrying a crew – the first ship would be fully fuelled and ready for a return trip. In turn, their two remaining ships begin to fuel themselves for the next crews, and so on. These fuel weight savings open up many options for transport of equipment to the Martian environs, and lighter craft can make the voyage faster, preserving the health of any human crew. Automation and fuel production on site would revolutionize many aspects of settlement construction and habitation. The concept is so intriguing that official NASA Mars studies have begun to incorporate aspects of it in their own studies.

The idea of an incremental buildup of infrastructure is a concept familiar to arctic explorers and civil engineers at bases across Antarctica. Additionally, the concept of caching supplies in the field is a proven one. In early arctic and Antarctic exploration, explorers cached supplies ahead of their final expeditions, enabling teams to travel farther while carrying less (see Chapter 1). As we have seen, supply caches storing fuel and food are critical even today to the operations in Antarctica at large. On Mars, supply caches will likely be established not only on the surface but also on the moons Phobos and Deimos.

Mars visionaries would love to deploy caches between the planets, but how can stores be placed in empty space between two worlds? Mars and Earth present moving targets, making the emplacement of supplies at specific locations impossible. But there is a way to set down a network of infrastructure between the Earth and Mars. Rather than caching supplies at strategic positions, resources can be settled into looping orbits that intersect the two worlds. The idea is known as "cycling."

Cycling space stations could circle the two planets in long, looping orbits, forming an interplanetary transport system linking Mars and Earth. Because of the orbital movement of Earth and Mars, Earth moves faster around the Sun. A spacecraft can take advantage of the Earth's routine of overtaking Mars once every two years. Using the gravitational kick from an Earth flyby, the ship will sail by Mars five months from home. The cycling station will not stop but rather coast by, dropping off passengers in smaller ships. A second cycler can be timed to pass by Mars on its way back toward Earth, making the return trip in about five months as well. Opportunities for departures from both Mars and Earth arise every 20 to 30 months. Flight times between the planets vary with the amount of fuel used and the timing of departure, but if a ship is to follow this low-energy, free-return trajectory, the length of travel is locked down: roughly five months one way and 18 months the other.

Fig. 4.10. A cycling ship, built in a "Y" configuration to enable a slow spin for simulated gravity. This design, arrived at through Case for Mars studies, features conical ships for ferrying passengers to and from the surface of Earth or Mars while the cycling ship flies by (Painting by Michael Carroll, based on a study by Carter Emmart)

The cycling bases will provide crews with living quarters and vast storehouses compared to what could be carried on a direct flight, serving as moving way-stations. The cyclers can be large enough to spin, creating artificial gravity en route to their cosmic destinations.

Once supplies begin to accrue at Mars and crews start to arrive, the business of life gets underway. How would we build a livable environment on the red planet? Mars Society and other studies suggest that habitats should be housed mostly underground. A layer of between one to two meters of dirt can serve as a natural barrier to the ambient radiation on the Martian surface. Empty lava tubes, spotted on several Martian volcanoes, would be ideal locations. Habitats could be fabricated with the use of inflatable structures, erected hard shells, or even bricks fired from local clays. Mars rovers have detected the basic elements of clays, and if these are formed into bricks, Roman arch-style vaults would accommodate all necessary interior equipment and living spaces. The bricks themselves would need to be sealed so that interior atmosphere cannot leak out through the porous material.

Fig. 4.11. (*Left*)The early stages of a Mars base include solar power and buried structures for habitation. This arrangement looks quite similar to the layout of the area around the Lower Erebus Hut. (*Right*) Greenhouses on the Martian surface will need protection against radiation, low night-time temperatures, and occasional dust storms; accordion-style shields like these might be one solution (Paintings by Michael Carroll)

Other advanced studies have been carried out by the Four Frontiers Corporation in Florida. "Four Frontiers" refers to the Earth, Mars, the Moon, and asteroids as founts of natural resources. Founded in 2005, the Four Frontiers Corporation bills itself as a "space technology, entertainment, and education company." The group has hosted technical brainstorming sessions and conducts ongoing engineering studies, resulting in several detailed scenarios for setting down Mars bases. As in the Mars Society scenarios, most inhabited structures are at least partially buried for radiation protection. In the Four Frontiers plan, the vast majority of structures for human habitation would be encased within a butte or canyon wall. Sunlight would be shunted to the interior through spherical light reflectors installed on the outer slope of the natural landforms. The baseline study utilized a mesa in eastern Candor Chasma, a tributary of the great Valles Marineris canyon system that stretches across the Martian equator. In the Four Frontiers study, structures radiate out from the natural slope like the spokes of a wheel. Garages for rovers or other vehicles could be open shelters or sealable pressure chambers. Apartments would face out, affording radiation-hardened windows so the inhabitants can look out into the Martian wilderness.

Greenhouses would include a mix of surface and blanketed structures. Those built aboveground would lack the protection of soil but could use clear bladders of water instead. Water makes an efficient radiation barrier.

A gas exchange facility would be one of the first structures erected, manufacturing a host of gases that could be used for the production of various commodities as well as oxygen for the human occupants. Power from solar and/or wind power generators would supplement a small nuclear power plant assembled at a distance.

Fig. 4.12. The "Generation I" study carried out by the 4Frontiers Corporation included a site selection in East Candor Chasma (*Lower Left*), detailed floor plans (*Upper Left*) and renderings of the site as it would look in full operation (Images courtesy 4Frontiers Corporation, background painting by Michael Carroll)

As the designers of McMurdo have come to realize, recreation is critical to a healthy environment. McMurdo's pubs, library, and the Gerbil Gym help to keep morale up in dark winter months, while hiking and skiing trails in the summer season offer healthy activities to reduce stress and provide enjoyment of the Antarctic panoramas. Outside the station grounds, red-domed shelters along pathways, called "apples" afford survival sanctuaries should the weather turn bad quickly. Such survival shelters will be an important part of infrastructure for Martian travelers as well.

Interior spaces in the Four Frontiers Mars settlement include large indoor courts for sports, garden areas adjacent to the fish farms and gardens, and observation areas. Raised balconies provide high views of common areas and help to give a sense of expansive interior space.

McMurdo has some small greenhouses, and each is outfitted with seating areas. These green, leafy spots are very popular in the winter months. Greenhouses will be critical on Mars for multiple uses, including recreation and food and oxygen production. Four Frontiers agricultural expert Nathan Owen-Going suggests that each hardworking Mars explorer will require 2,975 grams of crops from 72 sq-meters of greenhouse space, allowing a daily energy intake equivalent to that used by farmers or active US military field personnel. A typical greenhouse in the Four Frontiers design covers 200 sq-meters, so one entire greenhouse will be required for every three inhabitants. Strategic concerns for Mars greenhouses include biodiversity, safety from solar radiation, and energy usage. Designers recommend LED lighting because of its low mass, long life, and low power usage. Additionally, LEDs can be adjusted to maximize the spectrum of light needed for photosynthesis.

Naturally, food must be brought to the red planet, and this will come in the form of prepackaged provisions, seeds for crops, and livestock for products like milk, cheese, eggs, and meats. Mars Society studies recommend goats for milk and meat. Tilapia is also a good food source, as the fish is nutritious, grows quickly, and does not require a great deal of space to survive. One concept calls for the tilapia farm to be combined with waterways in the greenhouses to create a more natural environment for humans and fish alike.

Designers envision bamboo as a critical crop. It grows quickly and constitutes a strong building material. The manufacture of concrete and steel on site is preferable to shipping materials from Earth, but metals require large-mass equipment. Scaled-down machinery will need to be shipped from Earth, resulting in the manufacture of narrower sheets of metal and smaller diameter pipes, rebar, etc. As the Four Frontiers study put it, Mars settlers "must strive to replace bulk strength with alternative, smarter design at an architectural level."

The key to any successful off-world settlement will be the ability to live off the land, making use of *in situ* resources whenever possible. We have seen that bricks can be fabricated from Martian clays. Masonry may become an important part of the mix: one resource that Mars has in abundance is rocks, and as a resource, these can be used in construction, the building of barriers, and the fabrication of roadways. Antarctica's bases of operation also utilize natural rocks. A rock factory has been established on the outskirts of McMurdo for similar purposes.

Lessons learned in Antarctica will provide valuable insights into future plans for a far more distant frontier than the harsh continent. The first Mars researchers and settlers will undoubtedly see echoes of McMurdo in their new home.

Planetary Protection

Protecting Antarctica's pristine environment is a high priority for USAP, and every visitor to McMurdo is briefed on care that must be taken to avoid contaminating the environment, including cleaning the soles of one's shoes and boots. For the same reason, vehicles must drive over metal grates before entering the vast ice plains of the Ross Ice Shelf where

Fig. 4.13. The rock processing factory at McMurdo Station (*Foreground*) sorts local resources by size (*Right*) (Photos by the authors)

aircraft come and go so that a minimum of dirt from inhabited areas is tracked into the unspoiled environment. We don't know yet if Mars has any life, but planetary protection is high on the list of concerns for planners. Planetary protection is the term given to the practice of protecting Solar System bodies (*i.e.*, planets, moons, comets, and asteroids) from contamination by Earth life, and protecting Earth from possible life forms that may be returned from other Solar System bodies. Despite the harsh Martian conditions, Mars' environment may be capable of supporting some types of terrestrial microbes. Extremophiles inhabit some of the most desolate and hostile places on Earth, and some would find Mars' radiation levels, low humidity, and frigid temperatures quite comfortable. It is imperative that humans prevent the introduction of Earth microbes to the red planet. The reasons are three-fold. Firstly, if microbial Mars life does exist today, germs brought from Earth could destroy Martian life. Secondly, any life inhabiting Mars might be detrimental to human or other terrestrial life. Thirdly, those searching for Martian life must be assured that any living organisms found on Mars are native and not hitchhikers from the spacesuits or rovers of Earth's explorers.

Preparing for Our Final Ascent

The delicate environment and harsh conditions of Antarctica provide a wealth of analogs for those who study future settlements on Mars, and they continue to challenge those who carry out Antarctic research today. When Ernest Shackleton advertised for explorers to accompany him on a South Polar expedition, his ad is said to have read, "Men wanted for hazardous journey. Low wages, bitter cold, long hours of complete darkness. Safe return doubtful. Honour and recognition in event of success." Today, Antarctic exploration still holds elements of danger (a well-loved researcher lost his life in a snowmobile crash shortly before we arrived), but as mentioned earlier in this book, the National Science Foundation and other organizations have set in place ways to make the Harsh Continent a safer place. Within this Mars-on-Earth settlement, we trained to prepare us for our work on Mt. Erebus. As one instructor put it, "Antarctica is going to try to kill you in lots of ways, and I'm going to help you avoid that." To that end, before we ventured up the slopes of Mount Erebus, the NSF sent us through a series of classes to prepare us for the experience. These classes were mandatory for all who travel to such remote and rugged areas on the continent. In fact, you are not allowed to venture outside the confines of McMurdo – even for recreational purposes – before taking a class.

Depending on the environment for which they are bound, researchers must submit to training in many aspects of polar survival. We were enrolled in a wide variety of classes, as we were headed to a volcano rarely visited by humans and the NSF wanted us to be prepared. We were about to attend McMurdo U, with curriculum covering high-altitude first aid, familiarization with snowmobiles and helicopters, winter camping, crevasse escape, glacier safety, the proper use of various radio equipment, and even techniques for cleaning up and prevention of environmental spills.

McMurdo U's Curriculum

Our "freshman initiation" class went by the innocuous name of "High Altitude Health." Since we were headed to the high country of the Harsh Continent, working at higher altitudes than most NSF grantees or scientists do, issues like altitude sickness and cold weather first aid were of paramount importance on our agenda. According to our NSF class welcome sheet, "This 2-hour course is designed to introduce altitude physiology, illness, prevention, and treatment information. This is required if you will be working above 8000 feet without close support."

Perhaps the most critical bit of trivia was the one our instructor opened with. He drew two pyramids on his whiteboard, labeled one "Equator" and the other "Antarctica," and then proceeded to scrawl a series of ascending numbers on the equatorial triangle. These moved from sea level to 12,000 feet (the scale we used in class). On the slope of the Antarctic triangle, the corresponding numbers began at the same value at sea level but rose more quickly. At its 9,000-foot altitude mark, air pressure in Antarctic is equivalent to the pressure at 10,000 feet on the equator. A 12,000-foot summit in Antarctica (such as Mount Erebus) experiences the same air pressure found on top of a Colorado "fourteener" (a 14,000-foot-high summit). The discrepancy occurs for two reasons. First, the spin of the planet thickens the atmosphere at the equator and thins it at the poles. Second, the air over Earth's poles is compacted in the cold temperatures. Air pressure at sea level in Antarctica is actually slightly higher than it is at more northern latitudes, but the pressure drops off more rapidly with altitude. This would be a significant factor in our Erebus experience.

Following this introduction, the teacher presented a series of diagrams and ratios showing what happens to a body's biology at higher altitudes. It's not a pretty picture. As the body's chemistry changes, the person undergoes aches and pains and a general feeling of malaise. Most personnel headed for any intense working conditions at altitude are prescribed a medication like Diamox to subdue the effects of the low-pressure environment.

When a person is not able to acclimate to high altitudes, several sicknesses can occur in increasing order of danger. The first is Acute Mountain Sickness (AMS), commonly experienced by those flying from sea level to the South Pole at 2,836 meters or Fang Glacier on Erebus at about 3,050 meters. Symptoms include headache, shortness of breath with any mild exertion, lack of appetite and energy, and insomnia. These usually resolve within 72 hours if the patient can rest and hydrate. A more dangerous disorder goes by the name of Severe AM, or High Altitude Cerebral Edema (HACE), and it happens when the brain swells and will not function normally. The symptoms can be similar to AMS but extend to a severe headache, loss of coordination (ataxia), and altered mental status. This can include disorientation, dizziness, and personality changes such as combativeness or general irritability. The worst cases of HACE result in unconsciousness and are considered a medical emergency.

The third and most critical of high altitude sicknesses is High Altitude Pulmonary Edema (HAPE). At this dangerous stage, excess fluid builds up in or around the lungs, cutting down the patient's ability to take in air. Symptoms include inability to catch one's breath even at rest, congestion and tightness in the chest, inability to carry out tasks or function normally, a cough that degrades into a wet, productive cough, pale coloring, extreme fatigue, profuse perspiration, fever, and rapid pulse rate. This constitutes a critical medical emergency. In all of these cases, treatment includes oxygen, the use of what is called a Gamow bag, and certain prescription medicines. Immediate descent or evacuation by aircraft is called for.

After surveying all the ways that our bodies might derail our high altitude adventures, our instructor unrolled a large orange bag with a small window in one end. This was a high altitude first-aid triage unit called a Gamow bag. The Gamow bag looks like a sleeping bag, but its zippers hermetically seal the interior. A foot pump is affixed to the lower end of the bag. The patient lies in the bag while a colleague pumps up the pressure or, if electricity is available, connects the pump's cable to an outlet. Care must be taken, as the patient's ears may hurt or "pop." Persons on the outside need to monitor the patient through the window to ensure that the patient does not succumb to claustrophobia. The bag can be pressurized to such an extent that the pressure inside is increased by the equivalent of a 1,000 to 4,000-foot descent (depending on the altitude of the injury). Oxygen can also be infused into the bag or given through a masked bottle, and a pressure relief valve can lower pressure if necessary. This portable hyperbaric chamber is lightweight enough that it comes in a backpack.

The Antarctic wilderness is relentless and uncaring. But spending hours with experts who are willing to prepare us for the worst emergencies gives researchers and grantees the confidence that, no matter what, someone in the know is watching out for them.

Fig. 4.14. Our classes at "McMurdo U" (*L* to *R*): NSF's Valentine Kass, tries out a high altitude Gamow pressure bag; crevasse escape training, indoors and outside; the ins and outs of helicopter travel: mountaineer Evan Miller and Rosaly Lopes prepare for liftoff (Photos by the authors)

Communications

Our next seminar took place in the two-story Building 165 at the McMurdo operations facility (MacOps). The place is a well-oiled machine and is the nerve center of the station, combining air traffic control, weather forecasting from satellite and surface station real-time data, communication between all camps and facilities, and coordination of fire, search-and-rescue, and other emergency personnel. It's a busy site. MacOps handles every kind of communication imaginable, from birth announcements sent from the US to emergency messages sent to the most distant of the field camps. The center must manage messages from helicopters flying to the polar deserts of the McMurdo Dry Valleys. They relay critical data as the large planes take researchers to the rugged Transantarctic Mountains, the isolated ice fields where meteorite hunts take place, and to the ships at sea. MacOps must deal with up to 36 field camps involving more than 330 personnel. The office even handles email traffic between camps and the station. MacOps serves as a source of important news, an occasional social outlet, an encourager, and a lifeline.

Within McMurdo, a phone system, several TV stations and internet serve the hundreds of inhabitants and visitors. In the field, Iridium satellite phones provide several channels: MacOps, MacWeather, Helo Ops, Medical, Search & Rescue, and Fixed Wing Ops. HF radios are available, while handheld VHF radios are more commonly used. Communication is sometimes circuitous. At the Lower Erebus Hut, for example, the encampment is on the aspect of the mountain opposite McMurdo, so line-of-sight communication is not possible. Instead, radio transmissions are relayed through a repeater on Mount Terror, which passes the signal on to McMurdo. On our expedition, our call sign is the project number given by NSF to us: W487. A typical transmission from the LEH begins with, "Mac Ops, Mac Ops, this is Whiskey 487 on Mount Terror. How copy?" On the other end, the transmission is rated on a scale of 1 to 5 for clarity. For shorthand, many operators use the term "Five by five" to indicate clear reception.

Antarctic Field Safety

The Field Safety class (also known as Survival School) is one of the most popular among grantees and McMurdo employees alike. It is a requirement of all United States Antarctic Program participants destined for the field. The class covers such subjects as the use of Extreme Cold Weather gear, frostbite and hypothermia, erecting tents, carbon monoxide poisoning prevention, and the operation of camp stoves. Attendees must learn to erect several types of tents and build emergency shelters as part of the cold weather and survival skills. The class allows field team members to work as a cohesive unit before going into the wilderness.

Any person slated to leave the established road system in McMurdo and its surroundings must complete training. Courses are tailored to the needs of each group. While some students learn sea ice protocols, our team was destined for the high country, so we concentrated on glaciers and high altitude subjects.

Fig. 4.15. Our telecommunications briefing at MacOps (Photo courtesy Elaine Hood)

Glacier Travel and Crevasse Rescue

In order for us to safely reach the ice column formations on the flanks of Erebus, the NSF assigned us a trained field safety mountaineer, Evan Miller. An experienced contractor from Washington State, Miller became one of our instructors, ushering us through our FSToP training (the USAP's Field Support Training Program). Additionally, Miller advised us about assembling our supplies for the trek before us, including boots, crampons, camp stoves, and food supplies.

Like many of Antarctica's expert guides, Miller's expertise was hard-won. Growing up in New Hampshire, he spent much of his childhood backpacking with his father and brother. Miller went on to law school and became an attorney in Cambridge, but the wilderness called to him. Thereafter, he trained as a mountaineer and climbing instructor, eventually ending up in the famous nonprofit National Outdoor Leadership School (NOLS). Over the course of the ensuing years, Miller spent 150 weeks in the field, leading clients to Alaskan glaciers, two-week ski excursions in the wilds of Wyoming, and climbs to Himalayan camps. His years of experience prepared him well for the Antarctic wilderness.

Miller took us through diagrams of how crevasses form, what to watch for, and what kind of emergency equipment we would have in the field. While our plans did not call for traversing much glacial terrain where crevasses are prevalent, we would spend some time on Fang Glacier. We would also be exploring areas beneath which are volcanically generated ice caves. These underground vaults are capped by ice ceilings that may be unstable or too thin to support a person's weight. Escaping from a cave fall requires techniques similar to those used for extricating oneself from a crevasse.

Miller began by showing us the ropes, literally. He taught us how to tie knots and shimmy up a single line should we find ourselves dangling from a rope in a hundred-foot ice fissure. We then climbed into a bright-red Haaglunds to track out into the Antarctic ice fields. Our route took us uphill out of McMurdo on the crushed rock roadway, down a slope past the New Zealand Scott Base some two miles away, and on to an overlook above the Ross Ice Shelf. In the distance, across the Ross Ice Shelf to our right, the mobile buildings of Williams Field stretched in a multicolored line along the brilliant horizon. A Twin Otter rose into the sky, the drone of its engine breaking the chilled silence of the white world before us. To our left ahead, the great cone of Mount Erebus trailed a banner of steam across the blue Antarctic sky. Miller climbed from the Haaglunds, brushed off the sides and wheeled tracks, and then drove us over a rough grate to knock off any detritus from the McMurdo environment. Our newly cleaned vehicle proceeded along the shoreline to a rise dotted with a train of black flags.

Our training group consisted of Rosaly Lopes, Michael Carroll, and Valentine Kass, Director for the National Science Foundation's Artists and Writers program. After tying us together in a line, Miller led us up a slick incline toward the ominous black flags. The crevasse on this hillside was a well-used training site, but it was dangerous, changing and shifting each year at the whim of the weather conditions. Our leader probed the ice with a long black-and-white pole, mapping out where the crevasse lay hidden beneath snow and ice. We could not have seen it and were surprised when the long pole went deep into the snow. This is the danger of crevasses – that they are so camouflaged. Miller and others experienced in these conditions can assess whether a terrain is likely to have crevasses and where by looking at small variations in slope. After some time poking the crevasse with the pole, he decided that because of the way the ice had crowned the crevasse, a descent would not be safe for us. Instead, we retreated downslope to a huge ditch dug out of the sea ice by earth-moving equipment. This artificial crevasse served as our training site. We secured various lines through a series of carabiners, simulating an emergency rescue of a fallen comrade. Our victim was a backpack, which we hoisted into the tame abyss. From inside, the trench showed a cross section of the layers of ice, folded and undulating as years of seasons deposited new blankets onto old to build the Ross Ice Shelf. The trench walls were a time machine, telling stories of ages past in subtle lines of blue, gray, and brown.

After nearly a week of focused training under Miller's wing, we dutifully swallowed our Diamox altitude medication and helicoptered to our first stop: Fang Glacier (see Chapter 6). The high-flying glacier is a required two-night stop for acclimatization before heading to the summit of Erebus. Fang is a sheet of polished ice resting precariously on a rocky shoulder just 3,000 feet below the volcano's crest, and it served as our gateway to higher ground. The Lower Erebus Hut awaited us, only one of many field camps scattered across Antarctica.

5

Working on the Edge

Antarctica provides scientists with excellent research opportunities in a wide range of topics, including astronomy, atmospheric and climate sciences, marine biology and eco-systems, and earth and oceanic sciences. While we were focused on geology and volcanol-ogy, we had the amazing opportunity to meet others who were engaged in fascinating field and laboratory work. McMurdo's "science edifice," the Crary Lab, contains laboratories that support many different types of research projects and provides limited office space for scientists. It is both educational and inspirational to get to know the different research teams at McMurdo. Many of these scientists go out to remote field sites, sometimes for day trips but more often for weeks or even months. It all depends on individual research needs and availability of field sites, availability of assets, and access to transportation. Before the summer field season starts, the NSF has to carefully evaluate resources and schedules. This includes support personnel such as mountaineers and cooks, and modes of transportation ranging from planes and helicopters to trucks and individual snow machines (skidoos). Problems can have a domino effect, and weather is always an unpredictable factor. Fuel has to be delivered to remote field sites and, if the weather is bad or there are mechanical problems with planes, delivery gets delayed and no personnel can be there. Even getting to the Amundsen-Scott South Pole Station can be problematic due to weather. While we were at McMurdo, a fellow grantee from the Artists and Writers Program had waited for two weeks for his flight to the Pole. Finally, the weather cleared and he boarded a flight. At lunch that day, we commented about how great it was that he was finally able to go, only to have him join us partway through lunch – his flight was "boomeranged" due to changes in weather conditions.

During our stay, personnel were trying hard to cope with the smaller than usual number of LC-130s. Those planes are aging, and mechanical problems kept a couple on the ground, significantly impacting operations.

Teams planning to go to remote sites understand that patience is needed and try to stay upbeat. The romantically named Yesterday Camp (see Chapter 4) is one of the several remote field camps operated by the US Antarctica Program during the summer season. There are seven main field sites that have resident staff to provide logistical and

© Springer International Publishing AG, part of Springer Nature 2019
M. Carroll, R. Lopes, *Antarctica: Earth's Own Ice World*, Springer Praxis Books,
https://doi.org/10.1007/978-3-319-74624-1_5

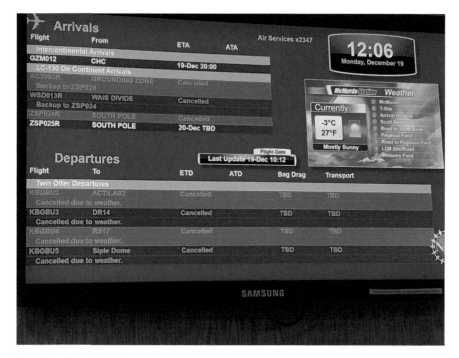

Fig. 5.1. Cancelled flights appear in red on a master monitor just outside the galley at McMurdo. (Photo by the authors)

operational support to researchers. Yesterday Camp is one of 60-plus smaller field camps, which include the Scott tents on Fang Glacier, the staging site for going up Erebus.

The seven major field sites are named Dry Valleys, Marble Point, Siple Dome, WAIS Divide Field Camp, Byrd Camp, Shackleton Camp, and James Ross Island Camp. Some support research projects while others are part of the logistics network.

Marble Point supports refueling operations for helicopters working in the Dry Valleys and on local sea ice. This remote camp is situated on a narrow strip of land between Wilson Piedmont Glacier and the Ross Sea. The camp is operated by a few full-time staffers plus fuels operators (called "fuelies") that rotate between there and other locations. Located only 46 nautical miles from McMurdo, it is close enough that fuel and equipment can be delivered by land traverse as well as from the sea via icebreakers.

Siple Dome is primarily a fueling point for aircraft. It is dedicated for those teams flying between McMurdo and West Antarctica or the South Pole, located 507 nautical miles from McMurdo. The staff provides crucial daily weather information to ensure flying is safe. Some researchers have conducted science operations out of Siple, such as Washington University's deployment of seismometers across the Ross Ice Shelf during the 2015-16 season. The name of the camp comes from the fact that it is located atop a dome about 100 km across near the Siple Coast along the east side of the Ross Ice Shelf. It is named after Paul Allman Siple, an American geographer and member of Admiral Byrd's expeditions. He is best known for coining and – along with colleague Charles Passel – deriving the first formula for wind chill factor. Anyone working in Antarctica today knows the importance of this term.

Byrd Camp, named after Admiral Byrd, is the third of the major field sites that, these days, primarily exists for operational support. Formally called Byrd Station, it was established by the US and originally served as a research station during the International Geophysical Year in 1957. The location, 1290 kilometers from McMurdo, is in the middle of the West Antarctic Ice Sheet, which is one of the fastest-warming places on Earth. Byrd Camp supports operations in West Antarctica and has a skiway that backs research flights such as radar-equipped airplanes that have obtained data from underneath the continent's fastest-moving glacier at Pine Island.

Shackleton Camp, sometimes known as Shackleton Glacier Camp, is located 850 kilometers from McMurdo. The camp supports science groups working in and around the Transantarctic Mountains. It is appropriately named after Sir Ernest Shackleton, who led the *Imperial Trans-Antarctic Expedition* from 1914 to 1917 (see Chapter 1), more often known as the *Endurance Expedition* after the ship's name. This was the last major expedition of the Heroic Age of Antarctic Exploration and, although it failed in its main purpose of making the first land crossing of the continent, it became widely recognized as one of the great feats of exploration. The camp is located near the Shackleton and McGregor Glaciers and also near bedrock outcrops that have produced important fossil records. The site was chosen in 2006 as optimal for research into paleoenvironments, landscape and glacial evolution, and ecosystems.

The James Ross Island Camp, located in the Antarctica Peninsula close to South America and 305 kilometers from McMurdo, has often supported ship-based researchers who field excursions to places of interest in the area of the James Ross Basin. The island was connected to the Antarctic mainland by an ice shelf until 1995, when the ice shelf collapsed. The island is an excellent location for studying biology and paleontology. For example, analyses of fossils collected there can provide understanding of how the Antarctica peninsula contributed to the dispersal of species between West Antarctica and southernmost South America at the end of the Mesozoic and the beginning of the Cenozoic eras. In fact, Antarctica at large has grudgingly given up a wide variety of samples from its fossil record. During the season we were there, a team from the University of Queensland uncovered ammonites, fish, birds, marine reptiles like plesiosaurs and mosasaurs, and several dinosaur species from the Sandwich Bluff area of Vega Island (adjacent to James Ross Island). These discoveries came primarily from the Mesozoic period (the age of the dinosaurs), but Antarctic fossils have been found dating from the Paleozoic (520 million years ago) all the way up to the Neogene (beginning 23 million years ago). These include wood, pines, ferns, ginkgoes, Synapsids (mammal-like reptiles) and other flora and fauna. The Antarctic Peninsula, popular these days with tourist cruise ships, is considered by many to be the most picturesque region of Antarctica, rich in marine and bird life.

Fig. 5.2. (*Left*) Rocky outcrops like this one in the Transantarctic Mountains near Wahl glacier display valuable sections of the fossil record (Photo by Peter Rejcek, courtesy NSF photo library); (*Right*) Paleontologists have unearthed fossil plants—this one from Oliver Bluffs near Beardmore Glacier—in rocks from a variety of Antarctic locations (Photo by Steve Roof, courtesy NSF photo library)

While at McMurdo, the WAIS Divide field Camp became a familiar name to us. Located 1,434 kilometers from McMurdo, WAIS Divide is a large camp with typically over a dozen resident staffers during the season. Land traverses take place between here and Byrd Camp, shuttling equipment as needed, while LC-130 and Twin Otter planes land by the camp to bring researchers and supplies. WAIS stands for West Antarctica Ice Sheet, and the WAIS Divide is the boundary separating two regions, one where the ice flows towards the Ross Sea and the other where the ice flows towards the Weddell Sea. The WAIS itself is the segment of the continental ice sheet that covers West Antarctica, also known as Lesser Antarctica, located on the side of the Transantarctic Mountains that lies in the Western Hemisphere.

The WAIS Divide Camp can support several research teams during a season and has housed important projects, including the well-known West Antarctic Ice Sheet Divide ice core project. This landmark project investigated climate changes on Earth by collecting and analyzing an ice core. The project's goal was to obtain records of the concentration of greenhouse gases in the atmosphere and of the Antarctic climate during roughly the last 80,000 years with the highest possible temporal resolution. The site of the coring was also called WAIS Divide, though of course the location is quite specific to the drill site. The site was selected because the snow that falls at WAIS Divide rarely melts, thereby building up thick annual layers that are compressed into ice by subsequent snowfall.

The coring happened during several field seasons, starting during the summer season in 2006-2007 and ending in 2011 after reaching a depth of 3,405 m (11,171 feet; over 2 miles), recovering the longest US ice core to date from the Polar Regions. The coring was stopped about 50 meters above the bottom of the ice sheet to leave a barrier between the borehole and the pristine aqueous basal environment. Coring was carried out using a deep ice sheet coring drill developed and operated by the University of Wisconsin in Madison. Subsequent analysis of the core showed that the ice at the bottom had fallen as snow 67,748 years ago. The age of the ice is determined by identifying chemical and physical differences between winter snow and summer snow, then counting the years in a similar way to how botanists use tree rings to work out the age of a tree. Analysis of the ice from many thousands of years ago contain precious information about climate conditions when the snow fell.

The 122-millimeter (4.8-inch) diameter cylinders of ice that make up the ice core contain uniquely detailed information on past environmental conditions, including the atmospheric concentration of greenhouse gases such as carbon dioxide and methane, surface air temperature, wind patterns, the extent of sea ice around Antarctica, and the average temperature of the ocean. This logistically ambitious and very successful project produced data that has been widely used in climate studies. The ice core, cut into 1-meter-long pieces, is stored at the US National Ice Core Laboratory in Denver, Colorado.

WAIS Divide Camp has played a key role in numerous other research projects. The POLENET (Polar Earth Observing Network) project is a global network dedicated to observing the Polar Regions. At the time of writing, during its second phase, POLENET deployed GPS and seismic instruments to advance understanding of geodynamic processes and their influence on the West Antarctic Ice Sheet. The United States Department of Energy has deployed instruments to measure climate conditions across the WAIS Divide. UNAVCO operates a GPS base station at the camp that can provide centimeter-level precision over a wide area when used in conjunction with geodetic quality roving GPS receivers. This camp is extremely busy every season.

The West Antarctica Ice Sheet is also of interest due to its geologic setting, as it overlies the ice-covered West Antarctica Rift System. In 2017, a team from the University of Edinburgh in Scotland identified 91 previously unknown volcanoes under the ice sheet using a variety of data, including ice-sheet bed-elevation data from which they located individual conical edifices protruding upwards into the ice across West Antarctica. Their work gave us the inventory of West Antarctica's subglacial volcanism and was a significant increase from the 47 volcanoes previously identified across the whole of West Antarctica, most of which are visible at the surface. Their results pose some important questions, such as whether there is still any subglacial activity today or whether activity could occur in the future. Evidence from studies elsewhere shows that removal of overburden pressure of the ice due to deglaciation can lead to increased volcanic activity and, therefore, increased melting of the ice.

If you ask McMurdo staff where they would most like to go, a common answer is the Dry Valleys, followed closely by Erebus. The Dry Valleys are the largest ice-free region on the continent. Located only 80 kilometers from McMurdo, they are a short 45-minute helicopter flight away, meaning that some research can be done as day field trips. Bell 212 helicopters are the workhorses for travel to and from the Dry Valleys. Helicopters typically fly 1,400 flight hours each season to various locations in support of US Antarctic programs.

Other teams need to stay for longer periods, either at the main base camp at Lake Hoare or at semi-permanent camps at three other locations. If summer research projects are expected to continue over several seasons at the same location, huts may be erected. These huts, which can be expected to last for several years, provide space, stable working areas, and comfort not achievable with tents. Huts have been used in recent years in Taylor Valley (an ice-free or dry valley in southern Victoria Land) for study of lake ecosystems, at Cape Crozier on Ross Island for population and behavioral studies of penguin rookeries, and

Fig. 5.3. A Bell 212 helicopter at the Lower Erebus Hut. These aircraft are the primary travel vessels to and from many remote camps, including those in the Dry Valleys (Photo by the authors)

near the summit of Mount Erebus for volcanology. Resupply and transport for the Dry Valleys camps happen by helicopter or tracked vehicle from McMurdo Station.

Science teams also erect small tent camps at various locations in the Dry Valleys. Every year, several groups conduct research in this fascinating region. As the name implies, the Dry Valleys are a polar desert, with precipitation (snow) being less than 5 cm a year. During the summer months, glaciers begin to melt, forming small streams that flow into the ice-covered lakes at the bottom of the valley. The Dry Valleys seemed devoid of life to Scott and his team who, in 1903, were the first humans to walk there. The inhospitality of the area left such an impression that they called it "Valleys of the Dead." Ironically, these days the valleys are one of the prime destinations on Earth for studies of life in extreme environments. There are plankton (a diverse collection of organisms living in water), rotifers (microscopic or near-microscopic animals), tardigrades (water-dwelling, eight-legged, segmented micro-animals), and Archaea (single-cell microorganisms). Algae grows in cryoconite holes, which form when windblown sand grains are deposited on top of the white surface of a glacier. The darker sand grains absorb heat from the sun, causing a small amount of ice to melt and form a pool. Organisms attached to the sand grains can grow and thrive in the water, forming algae. The Dry Valleys are a biologist's paradise. There are strikingly colorful bacterial mats growing at the bottom of some streams, surprising given that for most of the year there is no water in the streams. The bacterial mats are freeze-dried for months looking like dried sponges but, within 20 minutes of receiving water, they are re-hydrated. Researchers have been able to re-hydrate some mats that had been freeze-dried for over 26,000 years! The survival of these bacterial mats is quite amazing and helps researchers understand how life can survive in harsh conditions on other planets and moons.

The most amazing place in the Dry Valleys – undoubtedly one of the most amazing on the whole continent – is the site of the Blood Falls. The brownish red falls come from Taylor Glacier, the coldest glacier on Earth containing constantly flowing water. The unusual formation was first found by Australian geologist Griffith Taylor in 1911, after whom the glacier – and in fact that particular valley – is named. Taylor and his team noticed that a river had stained an ice cliff dark red, reminding them of blood. Initially, the red color was thought to be due to some form of algae, but later studies showed that the Falls are caused by an outflow of saltwater containing iron oxide, flowing from Taylor Glacier onto the ice-covered surface of West Lake Bonney. When the iron oxide saltwater comes into contact with oxygen, the iron oxidizes (rusts) and becomes red, staining the water the color of blood. Thanks to research done by a team from the University of Alaska Fairbanks who used radio-echo sounding (RES) to map out features below the glacier, we now know that the saltwater source is a subglacial pool of unknown size overlain by about 400 m (1,300 ft) of ice several kilometers from its tiny outlet at Blood Falls. The research team calculated that the saltwater takes approximately 1.5 million years to reach the Blood Falls, making its way through fissures and channels in the glacier towards lower pressures. Because water supersaturated in salt freezes at a lower temperature than freshwater, the water is able to remain liquid.

Fig. 5.4. The Blood Falls pour from the terminus of Taylor Glacier in the Taylor Valley. Iron-rich brine gives the falls their distinctive appearance (Photo by Elizabeth Mockbee, courtesy NSF)

The Blood Falls contain microbes that can survive the extreme environment under ice and with no sunlight. They live off minerals in the water, using sulfate and iron to help them metabolize organic matter. Microbiologist Jill Mikucki from the University of Tennessee has pointed out that the now-inaccessible subglacial pool that feeds Blood Falls was sealed off 1.5 to 2 million years ago and is therefore a "time capsule" for the ancient microbial population on Earth. Studies of these microorganisms could shed light into how microbial ecosystems survived when, according to the "Snowball Earth" hypothesis, the Earth was entirely frozen over during the Proterozoic eon about 650 to 750 million years ago. Life is persistent and can adapt to extreme environments: it could exist in equally challenging environments on Mars, Europa, Enceladus and beyond.

The Dry Valleys have many sites of interest for those studying life in extreme environments and its astrobiological implications. While at McMurdo, we met Sarah Stewart Johnson and her team. Sarah, an Assistant Professor at Georgetown University, researches the evolution of planetary environments, particularly with regard to the search for life on Mars and how life and its traces persist in extremely harsh environments such as Antarctica. Despite recent advances, there are many questions that remain unanswered, such as, what survival strategies do cells employ when pushed to their limits? Sarah and her team were conducting day trips to the Dry Valleys to collect samples that would be analyzed in the field and in the lab back at McMurdo. They were investigating whether Antarctica's paleo-lakes (ancient lakes that are now dry) harbor "microbial seed banks," caches of viable microbes adapted to past paleoenvironments that could help transform our understanding of how cells survived over ancient timescales. The paleolake sites in the Dry Valleys provide a great opportunity for the team to investigate questions about the persistence of microbial life, as these sites are thought to have remained geologically stable for millions of years. This geologic stability, together with the geographic isolation of the Dry Valleys and a steady polar climate, mean that biological activity has also probably remained stable over the last one to two million years.

The sampling is hard work, requiring sterile techniques at the field site and, back in the lab, the extraction of DNA from the organic material. Excitingly, Sarah and her team were conducting the first-ever DNA sequencing in Antarctica. They had brought along a miniature, handheld DNA sequencer – called a MinION sequencer – from Oxford Nanopore Technologies. As DNA passes through an array of protein nanopores embedded in a special polymer membrane, the sequencer measures changes in ionic current (current affected by the nucleotides present). The software driving the sequencer operated from a Macintosh laptop. We were indeed privileged to see some of the team's lab work, marveling at how far science has come in Antarctica in the last century.

Because of the cold, ambient temperatures in the Dry Valleys (roughly 2°C), the researchers anticipated that the sequencer might not be able to maintain proper operating temperature when used in the field and that the laptop battery could lose charge more rapidly. To compensate for this power loss and to maintain sequencer performance, the temperature of the sequencer and laptop was regulated using hand warmers and insulating materials. During this complex process of juggling equipment and warmers, team members had to be careful to keep the air vents on the sequencer clear for its operation.

Fig. 5.5. Members of the Georgetown University team, seen here in Crary Lab, included (*L* to *R*): Principal Investigator Sarah Johnson, Scott Tighe and Dave Goerlitz; (*Right*) the portable DNA sequencer used to chart ancient organic material from the paleolakes (Photos by the authors)

Findings from the Georgetown University team's project will be important not only for cell biology and Antarctica microbiology but also for planetary science. We could imagine future explorers on a Mars base doing the same, analyzing organic material from Martian paleolakes. Will they find life? Will the DNA be similar to what we find on Earth?

Fig. 5.6. The Georgetown University team at work in the Taylor Valley (Photo courtesy Scott Tighe)

Field Work Under Antarctica's Ices

The work on paleolakes is fascinating, and so is the work on actual subglacial lakes in Antarctica. Astrobiologists and aerospace engineers view Antarctica as a proxy for future exploration of oceanic ice moons like Europa and Enceladus. NASA's *Astrobiology Science and Technology for Exploring Planets* (ASTEP) has funded a project called SIMPLE, a multiple-vehicle, multi-season endeavor designed to conduct an in-depth study of the McMurdo Ice Shelf. Principal Investigator Britney Schmidt of Georgia Tech is collaborating with researchers from the University of Illinois at Chicago, University of Texas, Stone Aerospace, Moss Landing Marine Laboratory, and the University of Nebraska, Lincoln, to explore and characterize the ice shelf, to understand how the ice and ocean interact and support marine life, and to complete research over the course of three field expeditions. Their work promises to increase our understanding of ice-ocean systems on other planets, particularly on Jupiter's moon Europa.

The team's first season in 2012 served as a test season for ice shelf operations. Season 2 was a much larger undertaking, using robotic submersibles to explore above and beneath the shelf. 2015/2016's season 3 culminated in the bulk of the science operations. The five-foot-long SCINI (Submersible Capable of under-Ice Navigation and Imaging) remotely operated vehicle, developed at Moss Landing Marine Laboratories in California, imaged the environment below the shelf. SCINI is primarily designed for identification of organisms inhabiting the sea floor to a depth of 300 m, and it investigated water properties and chemistry. A follow-on vehicle, SCINI-Deep, is currently under development.

A second component of SIMPLE was called *Icefin*, built at and by Georgia Tech for use with the SIMPLE project as well as future polar oceanographic work. *Icefin* is a tethered autonomous underwater vehicle that combines relatively small size and human portability with a large suite of instrumentation that characterizes the water column, ice, and benthic environments (regions at the bottom of the sea). *Icefin* is modular vehicle and can be reconfigured easily, as well as transported and deployed in the field with little trouble. It carries forward and up/down imaging and sonars, a CTD, ADCP, DVL, IMU, and bathymetry. *Icefin* can dive to 1,500 meters, performing 3-km-long surveys below the ice.

The third member of SIMPLE's underwater vehicles was called *Artemis*, a large hybrid autonomous underwater vehicle built by Stone Aerospace based on the earlier DEPTH-X vehicle. *Artemis* carries remote and *in situ* instruments for characterizing the water, ice, and any microbiology found within and below the ice. The craft performs up to 15-km-long gridded surveys at the ice-ocean interface and along the benthic interface, and it characterized in detail the environment below the ice shelf.

In addition to the submersibles, SIMPLE utilized the Airborne Geophysical Platform HiCARS-2 as a simulation of a future spacecraft in orbit at Europa. In addition to both HF and VHF ice-penetrating radars, the platform includes gravimetry, laser altimetry, magnetometry, and imaging. This platform is actively testing the radars being developed for the Europa Clipper, the NASA mission that will survey Jupiter's icy ocean moon.

Over the years, other investigations have continued beneath the ice of Antarctic lakes as well as the more permanent ice sheets covering the oceans and bays of the seventh continent. These explorations include both human and robotic elements and hearken from many countries and organizations.

The Antarctica Search for Meteorites (ANSMET)

ANSMET, a program funded by the NSF, has brought back more than 20,000 meteorites since it was established in 1976. Teams look for meteorites in the Transantarctic Mountains, the mountain range that divides East and West Antarctica. This area is preferred because it serves as a collection point for meteorites that fell on the high-altitude ice fields throughout Antarctica and, over centuries, slowly travelled "downhill" while embedded in flowing ice. The Transantarctic Mountains provide a natural barrier for the flowing ice, trapping some of the meteorites. Erosion of the ice by wind can expose the meteorites and, since the search is visual, the contrast between the dark meteorites and the white snow and ice makes them easy to spot relative to other locations on Earth. Unfortunately, the vast majority of these precious rocks end up sliding into the ocean before they are found. The ANSMET program has found many of the rare lunar and Martian meteorites, including the famous Allan Hills 84001 (ALH84001), a Martian meteorite that some researchers in 1996 claimed contained microscopic fossils of Martian bacteria. Most scientists did not accept the controversial claim, arguing that the "fossil" features could be explained without requiring the presence of life.

The ANSMET program has been crucial for the study of meteorites. Together with programs run in Antarctica by other countries, it provides a treasure trove for researchers and educators. Many meteorites that are found around the world end up privately owned and not properly curated.

Planetary scientist Jani Radebaugh, who has long collaborated with Rosaly Lopes on studies of extraterrestrial volcanoes and has volunteered three times with the ANSMET team, explained to us how the search is done. Although most in the four to ten-strong search team are meteorite scientists, there is sometimes room for other planetary scientists and volunteers such as Jani. Members have to be ready to spend more than a month camping out on the ice field, typically five to seven weeks. Every day that conditions allow, they ride individual snowmobiles spaced about 30 m (100 feet) apart, scanning the ice fields for the black rocks. When a meteorite is found, the position is recorded with a GPS, the meteorite is given an identification number (such as Allen Hills 84001) and it is placed into a sterile Teflon bag. The specimens are kept frozen all the way to the Meteorite Curation Facility located at NASA's Johnson Space Center in Houston, Texas. The field work can be grueling for the ANSMET team members, particularly when weather turns bad and they have to stay in their tents, sometimes for days. But the science haul is worth the sacrifices: in one season, a team will typically collect over 1,000 meteorites.

Fig. 5.7. Members of the 2016/17 ANSMET team carefully search for meteorites within the pristine Antarctic ices (Photo courtesy Jani Radebaugh/NSF)

Field Work in Antarctica: The Myths and the Reality

Antarctica is one of the most interesting places on Earth to do field work. The competition among scientists to be able to get to the white continent is fierce. The US National Science Foundation and the equivalent entities from other countries can only support a limited number of researchers. Most first-timers to the ice probably feel like we did, extremely excited to be granted this amazing opportunity while at the same time somewhat apprehensive. Even if you have done a lot of field work before, Antarctica is like nowhere else – the environment is always trying to kill you, while you are always trying not to harm it in any way.

These two themes are with you from the time you hear about your successful application to the time you get back home. Intensive training is required to for your own survival, and more is necessary to preserve the unique environment.

Your initial application to the NSF to access Antarctica needs to spell out every location you hope to get to, for how long, what field equipment you wish to bring, and what facilities at McMurdo or elsewhere you hope to be able to use. Once the application is successful, an implementer will work to develop a detailed field plan, which includes use of field camp sites if needed, and use of everything from helicopters to snow machines. Once you arrive at McMurdo, training becomes intensive, always following the themes of staying safe and keeping the environment pristine. Training begins with lectures in Christchurch at the Antarctica Terminal, which is also where the cold-weather gear is stored and distributed. Training even includes IT security and testing your laptop to ensure it is free from viruses (Chapter 3). You, of course, also have to be virus-free, and it is mandatory to provide proof of a flu shot or get one on the spot.

Once you land on the ice at Willy Field and get on the bus or other transportation to McMurdo, you will notice red and green flags parade in long lines, each about ten meters apart, to serve as a way back home to the station if a whiteout occurs. As previously mentioned, blue flags mark fuel lines, which snake across the surface of the ice. The black flags (indicating hazards such as thin sea ice or crevasses) are an immediate sign that danger can lurk even on the apparently smooth and pristine ice surface. You might also notice metal grates that vehicles drive over, reminding one of cattle grids. These are there for a good reason: to protect Antarctica's pristine environment, vehicles must drive over these grates before entering the vast ice plains of the Ross Ice Shelf, where aircraft come and go, so that a minimum of dirt from inhabited areas is tracked into the pristine environment.

The field camps scattered across the Harsh Continent range from low to high altitude, from coastal to inland, and from icy to rocky. Three Antarctic destinations engender awe and some degree of envy because of their remote and exotic natures: the McMurdo Dry Valleys, the South Pole Amundsen/Scott Station, and the Lower Erebus Hut. The LEH would serve as staging area for our primary objective: the famed ice towers of Mount Erebus.

Selected List of Antarctic Specially Protected Areas

Antarctic Specially Protected Areas, or ASPAs, encompass both natural and historical sites. Below is a partial list of areas currently protected by international agreement.

ASPA #	Site	Comments
101	Taylor Rookery Mac. Robertson	Largest Emperor Penguin nesting grounds on land
105	Beaufort Island	One of the primary bird breeding areas
106	Cape Hallett, Northern Victoria Island	Main area of study is prolific biology impacted by human activity
115	Lagotellerie Island, Graham Land	Diverse plants and animals typical of the southern Antarctic Peninsula, including the only two flowering plants on the continent
118	Mt Melbourne, Victoria Land	Geothermal site with rich botanical specimens
119	Davis Valley/Forlidas Pond	Transantarctic Mountain Range; pristine freshwater ecosystems
121	Cape Royds, Ross Island	Most southerly Adelie Penguin rookery
122	Arrival Heights, Hut Point Peninsula	Series of low hills on Hut Point Peninsula, sheltered from electromagnetic noise, used for atmospheric observations

ASPA #	Site	Comments
124	Cape Crozier	Lower eastern slopes of Mt Terror; rich bird and marine mammal populations
125	Fildes Peninsula, King George Island	Most extensive snow-free summer coast on King George Island, late Cretaceous to Eocene fossil record
126	Byers Peninsula, Livingston Island	West end of Livingston Island; Jurassic and Cretaceous sites, which can be linked to other southern continent fossil records
129	Rothera Point, Adelaide Island	SE corner of the Wright Peninsula, near the British Rothera Research Station. Used as a control site in reference to human impact in other areas
130	Tramway Ridge, Mt Erebus	200-m square area that includes most of the warm, open ground of lower Tramway ridge (the upper ridge is a prohibited zone). High-altitude plants associated with active fumaroles
131	Lake Fryxell, Taylor Valley	Among the richest algal and moss growth in south Victoria Land Dry Valleys; serves as reference site for the ecosystems of other dry valleys
137	NW White Island	Isolated population of Weddell Seals
140	Deception Island, Shetland Islands	Unique flora communities associated with geothermal sites; includes 11 subsites
141	Yukidori Valley, Lutzow-Holm Bay	Near Japan's Syowa Station, fellfield ecosystem
142	Svarthamaren	Largest inland seabird colony on Antarctica
148	Mount Flora, Hope Bay	Rich fossil and geologic area, protected because of easy accessibility
151	Lion's Rump, King George Island	Rich populations of Elephant and Fur Seals, Adelie, Chinstrap and Gentoo penguin nesting sites, various other birds, and rare native flowering plants. Largely undisturbed by humans
152	Western Bransfield Strait	Marine preserve off the coast of the South Shetlands
154	Botany Bay, Cape Geology, Victoria Land	Botanical site rich in mosses, lichen, and algae. Site also includes rock shelter and artifacts from British Antarctic Expedition of 1910-1913
155	Cape Evans, Ross Island	Significant archaeological/historic site of Scott's Terra Nova expedition, used later by Shackleton
156	Lewis Bay, Mt. Erebus	Site of the Air New Zealand crash, November 1979. Protected in remembrance of the victims
158	Hut Point, Ross Island	Built by Scott's Discovery expedition, the archeological site encompasses historic structures and relics associated with the Heroic Age. Site was used by several expeditions
162	Mawson's Huts George V Land	400-m section of coastline in George V Land. ASPA 162 includes four archeological hut sites: Main Hut, Transit Hut, Absolute Magnetic Hut, and Magnetograph House. Huts served as Mawson's base for the Australasian Antarctic Expedition.
164	Scullin & Murray Monoliths	Greatest concentration of seabirds in E. Antarctica. Glaciers from the Continental Plateau flow around the monoliths into calving termini
166	Port Martin, Adelie Island	Archeological site containing structures and artifacts from several French expeditions from 1948 to 1952

ASPA #	Site	Comments
168	Mount Harding, Grove Mountains	The region's unique geology provides a record of the evolution of the East Antarctic Ice Sheet
172	Blood Falls, Taylor Valley, McMurdo Dry Valleys	These bright-red waterfalls drain from beneath Taylor glacier; site is preserved for its extraordinary microbial colonies and unique geochemistry

Fig. 5.8. The ASPA at Arrival Heights includes electromagnetic monitoring equipment from several countries, including New Zealand's bright-green facility at left. Most of Arrival Heights proper is to the left off frame (Photo by the authors)

6

Our Voyage Up The Mountain

Fig. 6.1. Emerging from a mist of its own making, the summit of Mount Erebus is seen here from near the Lower Erebus Hut garage (Photo by the authors)

© Springer International Publishing AG, part of Springer Nature 2019
M. Carroll, R. Lopes, *Antarctica: Earth's Own Ice World*, Springer Praxis Books,
https://doi.org/10.1007/978-3-319-74624-1_6

The 2016/17 season was a busy one on Mount Erebus. Some 32 National Science Foundation personnel and grantees stayed on the summit for stints lasting more than eight hours. Many came to carry out scientific research. Others came to repair the structures, experiments, and huts that humans have boldly erected and that the mountain's elements continue to sunder. No tourists came. To visit Antarctica's steaming mountain, one must have a very good reason, and only personnel with specific designated duties are allowed upon its crown.

Of all the established remote field camps, perhaps the most famous and coveted is the Lower Erebus Hut. The landmark has taken on an almost hallowed mystique among those who live and work at McMurdo. And it is no wonder that Erebus is such a sought-after location to visit: the volcano looms over McMurdo, peering across the station like a potentate surveying her realm. Visible from nearly every point in McMurdo and Scott Base, the summit catches the first morning light and the last dying embers of sunset. Its prominence and affect are a testament to how aptly the mountain is named. In the mythology of the Greeks, Erebus was the primordial god of darkness, the son of Chaos, and the companion of Nyx, goddess of the night. His mists enshrouded the world, hiding it from the celestial light. Each evening, Nyx drew Erebus' darkness across the sky, ushering in the evening. Erebus' daughter Hemera dispersed the mists at dawn, bringing daylight. The Antarctic volcano seems to retell the Greek drama with each long season. The mountain itself has a somewhat complex provenance for its title; it is actually named for Shackleton's ship of the same name, while Mount Terror, an extinct volcano adjacent to Erebus, was named for its sister ship, the *H.M.S. Terror*.

Mount Erebus has a special place in the hearts of New Zealanders, and that place serves as a cautionary reminder of the volcano's uncompromising bent toward lethality. On November 28, 1979, Air New Zealand's commercial flight TE901 took off from Auckland on a regularly scheduled sightseeing flight over Antarctica. The popular tourist flights had been departing since 1977 aboard McDonnell Douglas DC-10s. Flight 901 slammed into Mount Erebus under sketchy weather conditions with low visibility. In one portion of the cockpit transcript, one pilot remarked about how difficult it was to tell the difference between "the ice and the clouds." The aircraft carried 237 passengers and 20 crew – none survived. The crash came to be known as the Mount Erebus Disaster. Throughout the 70s, a string of air disasters had occurred involving the DC-10, so suspicion first turned to the aircraft itself. An initial investigation deduced that the crash was due to pilot error, but public outcry led to another study carried out by the Royal Commission of Inquiry. This board uncovered telling information: namely, that a correction was made to the coordinates of the flight plan the night before the flight, but the crew was not informed of the change. Instead, they operated under guidelines given to them 19 days earlier. While the crew assumed their flight computer was taking the aircraft down a safe corridor over McMurdo Sound, the plane had in fact been diverted directly into the face of Erebus. The commission's final report accused the airline of putting forth a "litany of lies." The incident ultimately led to a restructuring of the airline's senior management.

The loss of Flight 901 impacted the national psyche of New Zealand. The search and recovery operations that followed, involving New Zealand and US personnel, was physically exhausting and emotionally traumatic. Bodies, catalogued personal items, and

Fig. 6.2. New Zealanders erected this beautiful chrome Koru up the hill from Scott Base in honor of those who lost their lives on Flight 901. (Photo courtesy New Zealand Press Association)

wreckage were removed from Ross Island over the course of many shipments via land and sea. Air New Zealand's Captain Gordon Vette, one of the key players in the investigation, took great care in researching and publicizing what is known as the *sector white-out phenomenon* that befell the crew in its last moments. His research continues to help prevent recurrences of this type of disorientation for flight crews. Commercial airlines no longer carry out flights over Erebus, and the mountain continues to command great respect by those who dare travel up its slopes. The crash site is now a protected area, but a monument to the lost stands today at a stony outcrop three km from the site of the crash. A second memorial, a chrome Maori koru,[1] was erected in 2011 just outside of New Zealand's Scott Base.

---

[1] A Maori symbol imitating the spiral of an unfurling fern.

A Toothy Affair

While high-profile Mt. Erebus is visible for all to see from McMurdo and Scott bases, getting to its summit is akin to arriving at Franz Kafka's Castle. For those who plan an extended stay, it is a circuitous voyage to the top. After the obligatory pre-qualifications, immunizations, travel papers, survival training and high-altitude first aid courses (see Chapter 4), the first stop is a two-night stay at lovely Fang glacier, courtesy of the National Science Foundation. The extended stay at Fang Encampment assures that most grantees and researchers will arrive at Erebus' summit camp, the Lower Erebus Hut (LEH), in stable physical condition. This is important: medical crises at high altitude are serious affairs. Fang helps safely cull any people who might not fare well at the higher altitudes and acclimatizes those who can go higher. Despite the Fang "stopover," in the past 20 years over a dozen people have had to be evacuated from the Lower Erebus Hut due to altitude sickness.

Fig. 6.3. The most efficient way up to Fang is by helicopter, where the views of the volcano are spectacular (Photo by the authors)

Fang Glacier rests on a rugged shoulder of Mt. Erebus, along the inside edge of the greater caldera area surrounding the central peak. The camp is a 20-minute helicopter flight from the McMurdo helipad. The site is on the face opposite to the slope facing McMurdo, so all communication must be relayed through a repeater station on nearby Mount Terror. Temperatures at Fang range from an average of -20°C in the summer to -50°C in winter.

The Fang Glacier Encampment consists of a quartet of forlorn Scott tents situated on the ice plain, exposed to the polar elements. Travelers stay in Scott tents with no heat, few comforts, and an incredible view. 2,760 meters below, the ocean plays along pristine white shores. Volcanic rock rises through the shimmering frozen ice plateau in craggy piles of deep brown. The distant Transantarctic mountain range of Antarctica's mainland hunkers along the horizon on one side, while the summit of the dead volcano Mount Terror rises on the other.

Fig. 6.4. Making our way toward the isolated Fang Glacier encampment (Photo courtesy Evan Miller)

The Scott tents are largely unchanged since their design a century ago, when Robert Falcon Scott designed them for his expeditions. The four-sided pyramids stand nearly three meters in height, with a separate base 8 feet square. At the apex of the pyramid, four drooping tubes serve as vents, creating a subtle flow of air through the tent and keeping the atmosphere fresh.

The tents are incredibly stable, weathering 200 kilometer/hour windstorms. The inner walls are one of the more clever features of the design. Scott tents are double-walled. Early explorers discovered that water vapor from exhaled breath and sweat gathered on the inside of the tents, where it would freeze. When an inhabitant bumped the wall or when the wind caused the fabric to flex, a powder of ice rained down on people and equipment. Scott solved this problem by lining his tents with a layer of fabric porous to water vapor. Moisture passes through it, condensing on the outer, windproof wall where it flakes off but remains sequestered in between. The double layers also help to insulate the interior.

The front door of the tent presents challenges, especially when carrying equipment or supplies. It effectively keeps the cold out, but it nearly keeps the people out as well. Built like the iris of a camera, the round door becomes a short tunnel as one struggles to enter or exit. Its fabric tends to bunch up around a person's shoulders, hips and feet. The opening is small and low, requiring one to exit on hands and knees. In Fang's rarified air, campers think twice before embarking on the physically exhausting process of entry or exit. Thank God for pee bottles.

The center of the pyramidal tent is high enough that warm air congregates there, creating the perfect space to hang socks and other items. It is also roomy enough to stand in when one is donning or doffing gear. For more storage, pockets are sewn into the walls. Like most camping situations, mountains of loose personal items and hiking gear pile up. Time is spent just trying to locate various key items, which can congregate under sleeping mats or behind storage boxes.

Three of Fang's tent quartet are reserved for storage or sleeping (two people to a tent, and even that is tight). The fourth shelter is the charming toilet tent, also called the pee-pee teepee. The arrangements are practical: a bucket topped by a Styrofoam seat for pooping, and a barrel with a funnel for peeing. As with all field camps, everyone carries their own pee bottles. The contents of these one-liter Nalgene jugs are dumped into the barrel and eventually flown back to McMurdo for disposal. Our project used the two tents in the center for our living spaces. The two novices slept in the tent by the supply tent. The one adjacent to the toilet tent became our cook tent/mess hall as well as mountaineer Evan Miller's sleeping tent. There, Evan cooked us meals, heated water for coffee or hot chocolate, and warmed our personal water bottles for use inside our sleeping bags. The "hot water bottles" were the only source of heat inside our own tent, aside from body heat, which raised interior temperatures to a balmy −9 to −7°C. We could not use the camp stove to heat the tent. Guidelines stipulate that if any open flame is in use inside the tent, the front door must be left open. There is history behind the rule: several years ago, two researchers were nearly overcome by carbon monoxide fumes when a stove either malfunctioned or was set improperly.

The interior of our tent let in enough light for daily living activities, as the Sun never set. But flashlights were required to read labels and find small objects. Despite double walls and body heat, the interior remained bitterly cold. Even after we had been closed inside for hours, snow still blanketed the tent floor like powdered sugar. When stored outside of sleeping bags, our disposable bath wipes became a frozen block. Nalgene bottles of drinking water hardened into popsicles, and any pee bottles left on the floor froze solid. We learned our lesson early on: anything that needs to remain thawed must be stored inside the sleeping bag—with us in it—or in the pocket of a coat or pants.

Tourist comforts were not the objective of our stay. The Fang encampment serves only one purpose: to acclimate those headed to higher ground. Acclimatization itself can be an uncomfortable process. Common symptoms include headache, fatigue, shortness of breath, even HAFEs (high altitude flatus expulsion, more commonly known as high-elevation farts). Once settled in at Fang, our duties were simple: rest, eat, drink lots of water, stay quiet and, in the event of HAFEs, keep the front door open. During our first 24 hours, Evan encouraged us to stay in our tents, keeping physical activity to a minimum. We were too cold to hold any kind of reading material with bare hands, and oxygen deprivation made conversation entertaining, if slightly frustrating. We napped a lot. The second day, Evan took us on a short hike to the edge of the glacier. The stony berm at Fang's edge marked the perimeter of Erebus' outer caldera, a great collar ringing the mountain's upper summit. A precipitous ledge dropped away toward the ocean below. Icebergs dotted the distant horizon like an armada of glowing white ships. A pinnacle of umber volcanic rock rose from the polished ground at the edge of the ice field, puncturing the bright sky with its talon-like silhouette. An ice crystal halo spread its spectral colors around the peak, a circular rainbow crowning the sun beyond the stone tower. Temperatures reached highs of roughly -4°F (-20°C). With calm air, the environment was comfortable thanks to our trusty Big Reds. But because everything is blanketed in ice, even gentle breezes can drop the experiential temperature by upwards of 20 degrees.

With our short hike and a good, hot dinner of macaroni and cheese under our belts, we slept better during the second night period. Our bodies were beginning to adapt to the alien environment.

Fig. 6.5. (*Left*) Three of our four Scott tents at Fang. Our tent is in the foreground. (*Right*) keeping cozy inside (Photos by the authors)

On to Erebus

After the acclimatization at Fang, travelers journey to the summit via snowmobile or helicopter. Evan Miller characterizes the trip via snowmobile as "sporty" – the NSF dispatched a Bell 212 helicopter to take us to our final destination, the Lower Erebus Hut.

The 800-meter ascent from Fang to LEH is a spectacular one. When a helicopter lands on the ice at Fang, all bags, parcels and equipment must be secured. The blast from the "prop wash" is equivalent to hurricane-force winds, triggering a blizzard of ice crystals and sand and scattering everything not tied or held down. Goggles and good covering (balaclava, hat, gloves) during entry and exit of an operating helicopter are a must to avoid eye injury and skin burns from flying debris. After dumping all equipment into the "bird," passengers climb up onto the flight deck, slide the door shut, strap in, and sit back for the minutes-long ascent. Erebus displays the majestic feel of Fujiyama, its rugged volcanic rock piercing the polished white ices and drifting snows of the mountain's flanks. Unlike the classic cone shape of Fuji, the mountain is truncated on the top, where the summit collapsed into the great caldera that collars the upper mountain today. Fang camp and glacier lie on the flanks outside of the caldera rim, while the Lower Erebus Hut resides within. At the center of this caldera, the mountain once again rises, with a dark, gritty summit topped by a complex crater.

Fig. 6.6. The rugged flank of Erebus, seen from the helicopter window (Photo by the authors)

Even the formidable Lower Erebus Hut complex looks tiny, dwarfed by the glistening Antarctic wilderness around it. LEH perches on a small plateau at an altitude of about 3,346 meters. The complex consists of two permanent buildings, a seasonal "Rack Tent" structure, an outhouse, and a colony of Scott and mountain tents pitched in a crescent upon the snowy lava flows surrounding LEH proper. Among the buildings, fuel drums, wooden pallets, snowmobiles, and crates round out the encampment.

Central to the remote camp is the Main Lower Erebus Hut, a wood and metal structure partially buried on its downhill side. The permanent LEH was built from 1992 to 1994. The roof of the Hut is at ground level in the downslope, outer direction, while a wall with windows faces uphill toward the volcano's pinnacle. A stairway heads from ground level down to the main door. There are no metallic railings; a wooden handrail assures that bare skin doesn't get stuck to the chilled surface.

Fig. 6.7. *(Left)* The Lower Erebus Hut was first erected as a Jamesway hut. This photo is ca 1978 (Courtesy William McIntosh); *(Right)* the Lower Erebus Hut today is approached from the roof line. It faces the summit of Erebus, behind in this view (Photo by the authors)

Next to LEH is a smaller structure simply referred to as the Garage. The Garage serves as a repair hangar for vehicles and equipment, a workshop, and a storage and staging area for experiments. It has indoor heating to enable workers to carry out complicated mechanical repairs. A bright-orange windsock mounted to the roof aids in helicopter landings, and it is in front of this building that our pilot chose his landing spot. As the main rotor wound down and the air blast died away, LEH personnel scampered to offload our supplies and passengers.

Before entering the Lower Erebus Hut, we unpacked our gear in the nearby Rack Tent. Although the main LEH is warmed, it is not an area used for sleeping (except in extreme weather emergencies). There simply is no room. Bedtime takes place in the surrounding tents. The Rack Tent, a semi-rigid structure with wooden, double-doored entryways, is primarily used for storage and staging of equipment, although we were given space in it for our gear and sleeping bags at night. The Rack has a small Kuma stove that raises ambient temperatures by a few degrees, but the tent remains below the freezing point at its far ends. With no electricity, the tent's interior relies primarily on skylights. At the time of year with the Sun up 24/7, the inside of the Rack Tent is always bright.

As soon as we stored our gear, we made our way back to the LEH. The Lower Erebus Hut's exterior is a weathered red in color, topped by a wood/metal roof. Inside the main outer door is a mudroom/airlock chamber where frozen food is kept. The inner door leads to the fully heated interior. A diesel generator supplies electricity, and satellite link offers Internet access for working scientists. A communications station manned by the camp managers coordinates air and surface traffic between LEH, other camps, and the home base at McMurdo. A room-length shelf runs below the windows. A sink is inset into its surface, along with a four-burner gas stove, oven, and food preparation areas. The "refrigerator" is the storage area under the sink at the end of the room along an outside wall (which remains fairly cold). Two long tables serve as the congregating area for the tired

Fig. 6.8. Two views of the Rack Tent. Notice the blue "airlock" entrance. The interior shot was taken at 10:30 pm, lit only by natural light filtering through the skylights (Photo by the authors)

workers who come in from the field. A water dispenser at the far end of the room is refilled daily with melted snow. Cabinets and shelves store food and snacks. A limited supply of alcohol is available, as is a virtually unlimited supply of chocolate. Far from a simple luxury, the calories feed the metabolisms of those working at altitude, where caloric intake must be increased to keep up with the environment. The body expends a lot of energy in keeping warm and in breathing the thin air.

As LEH is a closed environment in a dangerous wilderness, safety is of prime concern. One afternoon, we witnessed an exchange between a researcher and several camp personnel. Upon entering the Hut from a day of outdoor work, the researcher said that he thought he smelled "just a whiff of gas." Without hesitation or discussion, mountaineer Evan Miller lunged for the nearest window and popped it open. The Camp Manager on duty opened the side door. Only then, when fresh air was flowing into the room, did a discussion ensue about where any gas might be coming from. The culprit turned out to be a minute amount of gas released when a propane stove had been lit, but the fast reflexes of the camp veterans reflected the top-notch safety training that NSF puts them through.

Lower Erebus Hut functions as the operational brain center for the Mount Erebus Volcano Observatory (MEVO), administered by the New Mexico Institute of Mining. Its five automated geophysical observatories run year-round using power from solar panels, wind generators, and AGM batteries. The instrument packages rest at elevations of 2,100 to 3,700 meters altitude and can withstand temperatures ranging from -20 to -60°C. Each station has a seismometer, GPS, and various combinations of infrared radiometers, tiltmeters, temperature sensors, and meteorological instruments. Data is transmitted simultaneously to New Mexico Tech and McMurdo Station in digital packets. USAP supports a McMurdo technician who monitors the equipment throughout the year. MEVO also maintains a thermal camera on the crater rim; when power levels allow, data is streamed to the MEVO website.

LEH personnel help maintain the equipment that continually observes the volcano's activity. Senior research scientists, graduate students, and others monitor the mountain's gas emissions and operate camera systems that image in the infrared and visible spectrum. Of particular interest is the lava lake, one of only half a dozen such bodies in the world. Its level and gas emissions are closely watched for both scientific and safety reasons.

An army of scientists has carried out research on Erebus over the past several decades. The plethora of valuable data has come at a cost: support structures, old experiments, and abandoned science and engineering equipment pepper the landscape above LEH. Over the past several years, LEH has hosted a steady stream of construction workers tasked with the demolition and disposal of this debris. One such site, erected by a NASA team in 1992, supported an experimental robot called Dante. Engineers designed the robot to rappel down steep slopes. At Erebus, Dante was charged with sampling gases coming from the lava lake. The robot made it 24 feet down into the crater when its fiber optics communications cable kinked and broke in the sub-zero temperatures. The research team retrieved most of the equipment and the delicate robot for future tests. All that was left behind was some cable and a metal structure on the volcano's rim. Those objects added to the litany of seismometer support equipment, storage crates, and even parts of snowmobiles strewn across the slopes of Erebus. Renewed environmental awareness in recent years has spurred the effort to clean it up, and LEH supports the people who do it.

It used to be a shorter trip to reach the volcano's summit. In 1978, workers established the Upper Erebus Hut. Prior to that, researchers had to carry out their work from tent camps stationed at about the same location. Volcanologist William McIntosh, co-director of New Mexico's Geochronology Research Laboratory, remembers, "UEH was a one-room structure, basically an insulated plywood box, approximately 16′ × 24′, with a single door on the north end, and a small anteroom/storage room between the entrance door and the main room." Like Lower Erebus Hut, the UEH served primarily as a common space for living, cooking, and conducting science research, with personnel sleeping in adjacent tents. Small windows on the east, south, and west sides illuminated the interior. UEH was equipped with a cooking counter, an oil-burning Preway stove for heating, tables, chairs, a single bed, and two large oxygen bottles and regulators. UEH relied on portable radios brought up for the summer season from McMurdo.

Fig. 6.9. Two views of the Upper Erebus Hut as it appeared in December of 2010. The structure was destroyed by a storm in 2015 (Photo courtesy Nial Peters)

From 1978 to 1984, the hut was used every year for science, with a maximum population of as many as 12 people. Erebus' summit crater was a steep but quick 15 to 20 minute hike. The Upper Hut also provided access to the nearby fumaroles, ice caves, and towers. But the hut's location had one serious drawback, says McIntosh: "It spent many hours in shadow when the sun moved behind the crater each evening, making for cold nights in the tents. Sandblasting by blowing snow and dust frosted the plexiglass windows." That blowing snow also periodically sealed the front door, necessitating a digging session for those who came to call. But McIntosh adds, "For the most part, the Upper Hut was pretty warm and cozy once the heater was cranked up."

The status of UEH changed dramatically when, from September to December of 1984, five seismic stations on Erebus' slopes detected energetic Strombolian explosions. The blasts ejected a series of molten bombs up to 1,200 meters into the sky, plowing furrows into the ground across an extensive area. "The slopes looked like they were littered with giant black cow pies," says McIntosh, "some as large as seven meters in diameter. The official geologic name for these is 'cow-pie bombs.' Some bombs would twist in flight and come to rest looking like huge cruller pastries."

During this time, Erebus expert Phil Kyle took a helicopter up to the UEH. As he stepped out, he spotted a smoldering cow-pie lava bomb just a few meters from the wall of the hut. Turning on his heel, Kyle leapt back into the aircraft and asked the pilot to take off as quickly as his little rotors would allow. In fact, several lava bombs the size of compact cars sailed out of the summit crater and landed well beyond the hut. Refrigerator-sized incandescent blobs rocketed as far as a kilometer away from the crater rim. The UEH, it seemed, was within range of the sporadic explosions so common with Erebus. It was time to abandon the Hut.

Planners decided to erect a new hut on the outer caldera rim, two km down from the crater and well out of range of any volcanic fusillade. Two carpenters, McIntosh, and another geologist constructed the original Lower Erebus Hut in December 1984. The first building was a Jamesway, a Korean War-heritage mobile shelter made of curved wooden beams and insulated canvas. At the time, Erebus was still quite active. The Jamesway shipped in four sections, covering a 16-foot-square footprint. It was expanded further in 1985. The canvas blankets of the Jamesway suffered damage from plume acids, requiring frequent repairs. Eventually, the current Lower Erebus Hut structure took its place, along with the second permanent structure, the Garage. This structure has a rolling door for vehicle access.

Various researchers occasionally visited the old upper hut in the ensuing years. UEH was paid frequent day visits, used as a warming hut, and occasionally employed for overnight stays by various personnel during Erebus' more tranquil periods. The heater was kept operational and fueled, and Coleman stoves allowed visitors to make hot drinks. Oxygen bottles were also kept in working order in case of emergencies. There was no radio in the hut during this period, and no one stayed for extended amounts of time. In 2005, several years after Kyle's helicopter flight, researchers again visited the hut. They had to excavate the door from a meter of snow, but the interior was intact. It was a busy year for the volcano; the area was plastered with bombs, and it was deemed still too hazardous to use as a permanent outpost.

On the wall of the UEH, visitors wrote messages, documented events, and scrawled cartoons. One researcher wrote a quote from Robert William Service's poem *Call of the Wild*: "Have you gazed on naked grandeur where there's nothing else to gaze on…" The words are gone now, vanished along with the walls of UEH. The storied hut finally succumbed to the elements during a storm early in the 2015/2016 austral summer, leaving only a trail of debris.

Fig. 6.10. The site of the Upper Erebus Hut today, now used as a snowmobile drop-off for those hiking to the rim. Note the circular scars at far right, relics of lava bomb impacts (Photo by the authors)

What infrastructure there is on the high slopes of Erebus, including the Lower Erebus Hut with its little tent village, would enable us to stage to our areas of research. Of primary importance in our search were the volcanic vents, caves, and ice towers because of their possible analogs on the moons of the giant planets along with a host of other planetary sites. We have seen cryovolcanism and its fingerprints only from a distance. What might these erupting formations look like at close range?

The Antarctica/Planetary Connection: Ground Truth

The cryovolcanoes that dot diverse sites like Titan, Triton, Enceladus and Pluto undoubtedly show a variety of forms. While the details of those cryovolcanic formations – and what they may look like on a human scale – are still speculative, the flanks of Mount

Fig. 6.11. The tent village surrounding Lower Erebus Hut. In the background, the large tower to the left is associated with Hut Cave. Note the green flags for navigation in a whiteout. (Photo by the authors)

Erebus may echo those more alien forms. Erebus interacts with ice in unique ways. Its sulfurous breath pours from vents on its flanks, blowing through the ices to form its towering edifices. These famous towers build above crystalline ice caves, one of our first stops on the Erebus agenda (see also Chapter 2).

A few hundred meters downslope from the Lower Erebus Hut lie several ice towers, rising above spectacular volcanically-heated caves. While warmer than the outside ambient temperatures, these caverns are far from molten lava tubes. They are just slightly warmer than the outside air – enough of a temperature difference to sculpt magical formations inside the ice, from complex stalactites to organic hollows.

We explored two caves, the Hut Cave and Helo Cave. Both are considered "dirty caves" in that they have been explored for decades by unprotected humans. Many caves higher up the mountain are pristine, containing rare microbial colonies. The delicate nature of these extremophile biomes requires that any explorer wear hermetically sealed suits to prevent the caverns' contamination.

We did not have permits to explore those exotic biological sites, but we were interested in the structures capping them and what they could tell us about cryovolcanic formations on Enceladus and other ice moons.

The subsurface chambers are invisible from above, and thin spots in their roofs pose a real danger to any unwitting traveler who might stray into their midst. Evan Miller led us in serpentine routes across this treacherous terrain, probing the surface to make sure it was strong enough to support our weight. The fall, he explained, would probably not kill us, but the Medevac would cost the NSF tens of thousands of dollars, use valuable limited resources, and put rescuers at risk as well. Falling through a cave ceiling would not be a prudent thing to do.

After a careful approach, we arrived at Hut Cave, so named because of its proximity to the Lower Erebus Hut. The cave is crowned by an impressive ten-meter-high pillar of ice, twisted and carved by countless blizzards and windstorms. The entrance was tight; Miller used his ice axe to dig out the edges, expanding it so that we could slide through in our bulky Big Reds. Our trusty mountaineer affixed several lines to aid in our descent into the shallow cave. After climbing down some six meters, our feet met ice-encrusted rock, a welcome solidity after the slick icescape above.

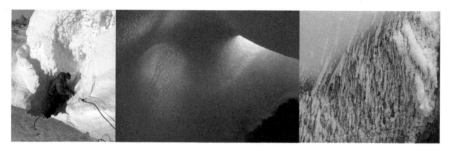

Fig. 6.12. Entrance (*Left*) and interior views of Hut Cave (Photos by the authors)

Climbing into the cramped caves was like visiting the animated movie set of *Frozen*. Light filters through the ceiling, painting the glistening walls, arches, and turrets in elegant shades of blue and turquois. Crystals grow on the surfaces of the rocks in the flowing air, pointing their lobate forms against the direction of currents. Neon cobalt cracks zigzag across the thin roof of the cave, confirming the structural weaknesses that we carefully avoided on approach.

Like the famous ice towers further up the mountain, Hut Cave feeds warm air from the bowels of the volcano toward the surface. Where that air meets ice, its flow gradually melts through the snow and ice, building a tower above. The interior was warmer than the temperatures outside, although not warm enough to cause moist puddles. Even the stone surfaces were dry to the touch.

The next morning, we were able to travel further to a more extensive cave called Helo Cave. The impressive site has developed multiple ice structures above it, great arched assemblies bowing as if in prayer toward the central grotto entrance. Another landmark accompanies the cave, an artificial one that lends the natural site its name: a crashed helicopter.

The helicopter does not mark another air disaster, nor did anybody die in the aircraft's demise. The event was in fact related to the same Artists & Writers program that we took part in. According to the writer Charles Neider, who was the grantee of the project and a passenger on the helicopter, the flight departed McMurdo just after noon on January 9, 1971. The helicopter's 4-person Coast Guard crew was tasked with taking Neider to Cape Bird, where he would camp with several New Zealand scientists researching penguins. The helicopter was heavily laden with supplies for the remote camp. Because of bad weather, the pilot did not follow the coast, which was customary, but rather headed toward

Fig. 6.13. (*Left*) The entrance to Helo Cave and (*Right*) its namesake, the site of a helicopter crash a few dozen meters away (Photos by the authors)

Erebus, perhaps trying for a shortcut. The craft climbed to 12,500 feet, just above the summit altitude of Erebus. With all its cargo, the helicopter was too heavy, and when a downdraft caught it, Neider and the crew crash-landed on the northwest side of the volcano. The rough touchdown occurred at 2:30 pm local time. The pilot transmitted a short Mayday call just before contact with the ground, but it was probably blocked by the mountain and was not picked up by McMurdo Ops. There were no injuries, but survival gear was at a minimum. The helicopter had been used primarily for local use around McMurdo, so survival bags had been removed. Fortunately, Neider had camping gear for his stay on Cape Bird, and there was plenty of food.

20 minutes after the chopper came to rest, the crew sent another emergency call out, and this one was received, triggering an immediate Search and Rescue operation. Helicopters and LC-130s scoured the coast along the normal route before eventually turning to the slopes of Erebus. At midnight, one of the Coast Guard's helicopters spotted the downed craft along with Neider's tent and several emergency flares. The crew dropped survival packs, and by 2:45 am a pair of Coast Guard helicopters was able to land at the crash site and retrieve the survivors.

The elements have not been kind to Coast Guard Helicopter #1404. Winds have blown it onto its side, and snows periodically cover almost all of it. But we could see its landing gear sticking up into the chilled air, its lanky rotor lying on the ground like a long arrow, pointing us in the direction of the famous Helo Cave.[2] The crystalline forms inside Helo Cave put to shame those we saw before. Here, aquamarine tendrils twist and intertwine down the surfaces of stalactites. The floor of the cave lies in a series of humps and hollows,

---

[2] The crash site itself is officially called "Helo Cliffs."

Fig. 6.14. Interior views of the magical Helo Cave (Photos by the authors)

soft curves of rigid ice and rock. Frozen domes hover over crystal-encrusted rock. Two kinds of crystals form here. The lava rock itself is infused with thousands of the famous Erebus crystals, and they, in turn, are encrusted by ice crystals in myriad forms: diamond-like, spaghetti strands, and organic-looking white worms slither across every face. Shimmering white light blazes from the entrance, but a softer blue takes over inside, especially where the ceiling is thin.

Caverns like these, in rock or ice, may well form subsurface labyrinths on other volcanic worlds like Io, Enceladus, Triton, and Mars. Several rock caves have actually been catalogued on the Moon and Mars. NASA's Lunar Reconnaissance Orbiter and Japan's Kaguya orbiter have identified over 200 holes and pits on many terrains across the lunar surface. These skylight openings appear to lead into subterranean lava tubes. Although different in formation from the Erebus ice caves, some analogies can be made.

We commonly see caves on empty lava tubes on terrestrial volcanoes such as those in Hawaii. Lava tubes result when low-viscosity lava flows cool and crust over, creating a roof above the still-flowing lava stream. When the eruption ceases, the duct empties, leaving behind an empty tunnel beneath the surface. More complex tubes form deeper within a mountain when lava between previous flows expands: the fluid pressure of the lava opens a channel that fans out in vast mazes of empty tunnels. "Skylights" commonly form when the ceiling of a subsurface lava tube collapses after it has drained of molten rock and then cooled.

Scientists who study the Moon have long known about collapsed lava tubes and channels that snake their way across its surface. Some have crumpled into long, winding valleys wandering across the lunar landscape. The formations, called sinuous rilles, are the result of the flow of lava. Astronauts visited one of these canyons, Hadley Rille, on Apollo 15. Networks of collapsed lava tubes draw webs across other areas. These have been found in many sites, including the Marius Hills region, the Sea of Serenity, the Sea of Tranquility, and even nestled in the Apennine Mountains. The lunar caves span diameters estimated to be from five meters, to over 900 meters long.

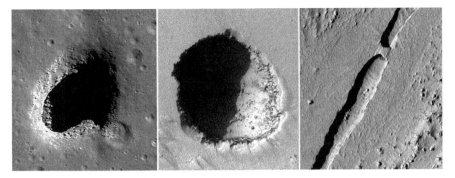

Fig. 6.15. Cosmic Caves (*Left*): A cave some 120 meters across marks a collapsed lava tube in Mare Ingenii on the Earth's Moon; (*Center*) a similar cave on the Martian volcano Pavonis Mons; (*Right*) a lava tube on Mars in a more progressed stage of collapse; here, only a bridge remains (Credits: JAXA Kaguya image, NASA/JPL Mars Reconnaissance Orbiter)

Openings above lava tubes have also been spotted in the volcanic provinces of Mars, such as Tharsis and Elysium. Like the caves on Erebus, Martian caves seem to be related to volcanic chambers or lava tubes. On the flanks of the great volcano Arsia Mons, observers have identified at least seven skylights formed by collapse. Several of the cave entrances would cover a football field, and some are so deep that orbital imaging systems cannot see the bottom.

The initial discovery came at the hands of NASA's Mars Odyssey spacecraft. Its cameras were not powerful enough to resolve details within the spots, but its thermal imaging system, which senses temperature, revealed that the dark spots gave off heat at night but were cooler than their surroundings during the day. This implied that these were not simple craters but rather holes in the ground. Later candidate sites appeared in images from ESA's Mars Express and the NASA Mars Global Surveyor, but the images could not clearly identify the nature of the dark ovals. Some researchers suggested that they were odd craters, while others proposed that they were pits or cavern entrances. Engineers called upon Mars Reconnaissance Orbiter with its advanced HiRISE camera system to investigate further. MRO's HiRISE (High Resolution Imaging Science Experiment) can pick out details as small as 25 cm per pixel. Flight controllers commanded Mars Reconnaissance Orbiter to scrutinize a spot on the volcano Arsia Mons. The powerful orbiter's snapshots revealed a scalloped edge encircling the entrance, and no visible walls. This told planetary geologists that the rim of the opening is smaller than the interior; the ceiling overhangs the chamber inside. It also must be quite deep, or image enhancement would have revealed interior details with just the ambient light from Mars' dusty atmosphere.

Several dozen such features have been found. Most appear to be associated with elongated depressions called pit craters, which commonly mark sections of a lava tube that has collapsed. While some entrances appear to lead into steep shafts, others are shallower. Mars investigators followed rows of skylights and pit craters on Arsia Mons for more than 50 km, finding that they led to outflow channels where the lava emerged when the tube was active.

Judging by the scale of the openings, Martian lava tubes may be gargantuan. Volcanologists have catalogued collapsed lava tubes in Hawaii and realized that the width and depth of a lava tube opening is in direct linear proportion to the size of the lava tube itself. If this ratio is similar for Martian lava tubes, the Martian counterparts may be two orders of magnitude larger. Subterranean tunnels hundreds of meters in diameter may be common. Like other features on Mars, their gigantic scale may be due to the lower gravity on Mars, along with environmental factors like temperatures and lower air pressure.

The caves found on the flanks of Arsia, Pavonis Mons, and at other sites may be prime candidates in our search for life on other worlds, as these are sheltered locations. On Erebus, we have seen how unique microbial communities spring up in Williams Cave, Harry's Dream, and other isolated sites on Erebus. The ubiquitous presence of microbes in extreme terrestrial environments causes optimism among many in the exobiology field that we may find biomes within the shelter of caves on other worlds. Earth's examples are varied and many.

The extremophiles on the sea floors of Earth's oceans are well known and serve as an archetypal example. Until 1977, sea floor volcanism was only a theory supported by circumstantial evidence. Oceanographers had noted hot undersea regions, but their nature and sources remained a mystery. That year, researchers using the deep-sea submersible Alvin discovered dramatic chimneys of sulfur compounds rising from the ocean floor along the Galapagos Rift zone near the Galapagos archipelago. In the years that followed, volcanologists came to realize that the number of volcanoes on the ocean floor must dwarf the 500 to 600 active ones on the surface. The majority of undersea hydrothermal vents occur along the mid-ocean ridges, where new crust is being created in the conveyor belt style of seafloor spreading unique to Earth. The mid-ocean ridge areas may well be the most volcanically active sites on Earth. Vents tend to cluster in groups, where seawater percolates through the Earth's crust, eventually making contact with hot magma below. Heated fluid makes its way up through fissures in the rock, leaching minerals along the way. When it finally streams into the ocean, it is laced with a complex mineral soup. In a maritime parallel of Erebus' ice columns, mineral-rich water erupts from these sources, building delicate structures of spires and chimneys, some of which may tower dozens of feet above the sea floor.

Perhaps the most remarkable phenomena associated with these submarine volcanic sites are the colonies of life huddling around them. Far from the sunlight that feeds Earth's complex web of life, the cold depths were thought to be a desolate, high-pressure desert brooding in eternal darkness. But the sulfur and other minerals carried by hydrothermal vents provide nourishment for an entire biome. Sulfur is taken up by bacteria that are completely independent of any food sources related to solar energy. Some of the bacteria associated with these vents are thought to be similar to the most ancient life on Earth and are known as archaebacteria. These microbes provide the foundation for an alien menagerie of frilled "Pompeii worms," blind crabs, giant tubeworms, one-eyed shrimp, and other strange beasts living in these frigid, high-pressure environments.

Microbial colonies have also been found within rocks kilometers from the Earth's surface. Lithoautotrophs ("Lithos" from "rock" and "troph" from "consume") obtain their life energy from minerals like sulfur and potassium. These microbes play an important part in turning solid rock into soil.

Enormous mine crystals in Chihuahua, Mexico harbor bizarre microbes that differ from any other known life on Earth. The giant calcium sulfate crystals grow in Mexico's Naica lead, silver, and zinc mine. Some of the beasts have been trapped within the crystals for between 10,000 and 50,000 years, remaining dormant the entire time. These robust little creatures live at depths of 100 to 400 meters below the surface, thriving in temperatures as high as 60°C.

Other extremophiles are halotrophic (living in high-salt environments), thermophyllic (tolerating high temperatures) or xerophyllic (found in dry environments like arctic or high-altitude deserts). Caves on Mars may provide an environment rich in minerals, with elevated temperatures, increased pressures, liquid water, and nurturing gases. The underground chambers could also offer shelter from the temperature swings of Mars' harsh environment, along with shielding from radiation.

Reaching our Goal at Tower Ridge

Exploring the caves was thrilling, but it was the structures above the subterranean underworld that held our attention the most. What insights could they give us about alien ice worlds? (See Chapter 7). After all, Erebus was one of the most alien places in our world. To get to the more developed ice columns, we had to circumscribe the mountain, traveling by snowmobile on a half-hour trek that took us halfway around and upslope by over a hundred meters. In some places, we had to lean in toward the hill to remain upright on the machines, with the steep slope of ice diving into a cloud bank below us.

In McMurdo, we had been given an overview of snowmobile driving and how to be a good passenger, and the lessons came to mind now. For example, when driving sideways across a steep, icy slope, our instructor suggested that we put both feet on the uphill side of the snowmobile and lean uphill. If the snowmobile should roll, you can avoid crushing your downhill leg under the vehicle. When driving down an icy slope like the one we traversed, the driver should never lock the brakes. This often leads to an uncontrolled skid. Instead, we were told to "embrace the speed." All things in moderation.

Although the slope we navigated consisted of ice and snow, the mountain continually reminded us of its volcanic nature. Black stones piled up like totems, staring down at us from spines of volcanic rock. Bastions of rock reared out of the snowy hillsides like ships cresting a wave, banners of vapor drifting from their prows. Here and there, the chilled breezes carried a sniff of sulfur to burn our throats.

The sun was nearly overpowering in its brilliance. The sky above us was pristine, the kind of deep blue only seen at high altitude. All the clouds gathered below us, blanketing the sea and coastline. A forest of communications antennas bristled from a rocky outcrop down the slope. We passed several areas marked with black flags. These were either dangerous to life and limb, or they were ASPA (Antarctic Specially Protected Area) sites, caves with preserved microbial biomes within. The ride in the cold air was chilly, but as soon as we stopped, we realized that the air around us was completely still. That was a good thing: breezes up here make working conditions miserable. Rounding a low ridge, we at last came upon them: The ice towers paraded in a line down the slope before us, following an underground chamber or channel. The Erebus colonnade marched in single file.

Many of the structures trailed steam in parallel lines to each other, looking like the smoke-stacks of an industrial-age British skyline. But this was not smoke: the undulating standards of vapor gleamed white against the rich cerulean sky. We pulled up on a gentle incline – the closest we could find to a flat parking spot – and secured the snowmobiles. Donning survival gear and hiking equipment, we headed for the glistening towers.

The Erebus parade varied in height: while a few of the fumarole crowns grew only a meter high, some of the pillars rose as tall as a five-story building. Each was unique, in constant change with shifting wind patterns and variation in hot air flow from within. Many rose in lobes and blades of translucent ice, sunlight flashing along sharp edges sculpted by Antarctica's relentless winds. Curved vents crowned many of the towers, surrounded by warped lips of frozen water. Many stood erect or corkscrewed to one side, displaying whimsical shapes like colonnades designed by a crazed gothic architect. Others seemed to have come from an artist more akin to Dr. Seuss, or Salvador Dali, with lop-sided protuberances and warped ice sombreros. Feathery white ribs gave way to rock-hard sapphire ices shimmering in the austral sun.

Fig. 6.16. The fumarolic towers could have been designed by a surrealist. The ice changes from season to season, so these specific formations will never be seen again (Photos by the authors)

In the field, frost formed within moments on our SLR camera, and the battery chilled beyond use within half an hour. We resorted to the fine little camera built into our iPhone for documentation. Were we seeing echoes of ancient Mars in a time when volcanoes ruled the atmosphere, erupting into a sky not so different from this one, with temperatures hovering near the freezing point? Certainly, Mars has had silicate volcanism much like we see on Erebus and other terrestrial volcanoes. Even Jupiter's "pizza moon" Io hosts active silicate volcanism, although the most spectacular of its eruptions are geyser-like plumes of sulfur.

Io is the quintessential volcanic world. With a diameter of 3,636 km, it is about the size of Earth's moon. Like the majority of moons in the Solar System including our own and the other Galileans (Europa, Ganymede and Callisto), Io is tidally locked into its orbit, meaning that the same hemisphere always faces toward Jupiter. Io orbits Jupiter at a distance of 421,800 km, circling deep within the planet's deadly radiation fields. Its day, equivalent to its orbital period, is 43 hours. And unlike the other gray-white moons in Jupiter's system, Io appears as a brilliant orange. Io also does not share the same frigid surface of the other three Galileans. Observers detect 100 times as much heat flowing from its surface as flows from Earth.

The reason, of course, is the moon's volcanoes. Io's surface is traumatized by three types of volcanoes. The first and most common of the volcanic wounds on the face of Io are lava lakes (see Chapter 2), in some ways resembling the one at the heart of Mount Erebus' summit crater. These collapsed sinkholes punch through the plains, leaving hot spots with temperatures as high as 1,300°C, and perhaps higher. The most classic example of these formations is an extensive basin called Loki (named after the Norse god of fire). Loki is a volcanic crater—or patera, as they are called on Io—filled with a lava lake some 200 km across (Fig. 2.7). Within its dark, viscous liquid is a gigantic island-like feature covered by frozen sulfur dioxide. The island is about the size of Rhode Island (most likely anchored to the rocky lake floor and made of silicates). In these lakes, cooled lava forms a crust that breaks up as it collides with the caldera wall, exposing a spider web of glowing lava cracks.

The second type of volcanic feature common on Io is lava flow. Molten rock and possibly sulfur has raced across nearly every square meter of the Dante-esque moon. The longest flow is called Amirani, which stretches across 300 km of rolling plains. Many flows are insulated, meaning that a crust of cooled lava covers them. These, are the sites that may contain caves like those on the Moon and Mars. An archetypal flow of this type issues from the volcano Prometheus. In the 17 years between encounters of the Voyager (1979) and Galileo spacecraft (whose first images of Io's surface arrived in 1996 and its last in 2001), the Prometheus plume moved over 80 km westward. Further analysis using infrared data showed that the volcanic source of Prometheus had not actually moved, but its lava flow had. As rivers of molten rock flowed across the landscape, the flows crusted over, probably creating lava tubes. About 80 km downstream from the vent, the lava broke out again, interacting with sulfur dioxide frost on the surface. This interaction is thought by some to trigger the visible plumes that erupt today. Alternatively, the plume may be caused by the hot lava eroding the surface and interacting with frozen sulfur dioxide underground.

The third type of volcanic feature on Io is the most dramatic. Explosive eruptions discharge the highest plumes seen anywhere in the Solar System. One example is the Tvashtar Catena region, where a curtain of glowing lava 22 km long fountains from a fracture in the floor of a great caldera. A nearby plume shoots gases over 100 km into the void.

Magma (molten rock) powers the eruptive activity of a volcano, but geysers operate quite differently. In a typical geyser, liquid accumulates in a vertical passage and is heated by underground sources. The liquid at the bottom of the shaft reaches the boiling point but is kept from boiling by the pressure of the liquid above it. Any slight change—a gas bubble or a particle of solid matter—can disturb the superheated liquid, causing it to explosively boil and vent up through the vertical duct. The definition of geyser on Earth is quite precise, as constrictions in the plumbing system are essential to the building up of pressure before an eruption. However, geyser-like activity occurs on Enceladus and likely on other icy bodies as well, so it may be far more ubiquitous through the Solar System than the more conventional volcanic eruptions seen on Earth and other rocky sites.

But can we stretch the point? Are the fumarole-fed towers of Erebus shadows of places on icy moons like Saturn's Enceladus or Uranus' Ariel or Miranda? We find flow features and possible vents on these tiny moons. But as Britney Schmidt warned, analogs can be tricky.

Enceladus Analog?

One of the most volcanically active worlds in our Solar System is Saturn's diminutive moon Enceladus. Positive identification of cryovolcanic activity on Enceladus came from flybys in February and July of 2005. The Cassini Imaging Science Subsystem (ISS) racked up a total of 377 high-resolution images of the Enceladan south pole. In several of these images, an average-sized house would be visible. Each successive flyby obtained more and more detail at progressively closer range. The south polar images showed huge grooves and ridges and no impact craters, indicating the surface is young. Folded, sinuous mountain chains, similar in scale to the Appalachian range in the US, bordered darker plains in the south. These plains were laced with organic material. Ridges and grooves deformed the topography, hinting at crustal slippage and deformation. The magnetometer detected ions streaming from the moon's rarified atmosphere.

The February flyby carried the craft over the surface at 1,167 kilometers. Data from Cassini's magnetometer discovered that something, likely an atmosphere, is pushing against Saturn's magnetic field around Enceladus. Professor Chris Russell from UCLA estimated that the moon loses about 125 kilograms of water to space every second. He explained that the loss is too great for the atmosphere to be formed simply by the Sun's photons bouncing off molecules from the moon's icy surface. In addition, Cassini's cosmic dust analyzer recorded thousands of hits from tiny particles of dust or ice, possibly coming from a cloud around the moon or from the adjacent E ring, a broad ring of dust-sized particles in which Enceladus orbits.

The results were so exciting and tantalizing that the science team made plans to return to Enceladus for a closer look. Detailed images obtained by Cassini's camera revealed icy swirls that appeared to have flowed. The ice movement was either similar to glaciers or perhaps remnants of very thick cryolavas. The case for some kind of active volcanism on Enceladus was building.

For the July 14 flyby, Cassini's orbit was modified, causing the craft to dive to within 175 km of the surface. The Cassini camera obtained detailed images of the south polar region of Enceladus that revealed a surprisingly youthful and complex terrain, almost entirely free of impact craters. The area is littered with house-sized boulders and tectonic patterns that are seen only in this region. Most exciting was that Cassini was able to fly directly through an extended plume of material. The spacecraft detected water vapor, carbon dioxide, methane, trace amounts of acetylene and propane, and possibly carbon monoxide and molecular nitrogen. Key evidence for active volcanism also came from the Composite Infrared Spectrometer instrument (CIRS), which detected enhanced temperatures in the "tiger stripes," fracture-like features in the south polar region. The tiny moon was constantly supplying material for Saturn's E-ring, a tenuous ring beyond those that can be seen through a telescope. Later, when spectacular images revealed geyser-like jets coming out of the tiger stripe fractures, all doubt as to what was happening on Enceladus melted away.

In 2006, researchers estimated that Enceladus loses 150 kg of water to space each second. While the material may not escape at a steady pace, the amount of water in Saturn's environment indicates the current level of activity has lasted for at least 15 years.

Just how similar might the Erebus formations be to the cryovolcanic vents in the rugged southern provinces of Enceladus? The geysers of the little ice moon erupt from a strange landscape. Multiple jets of icy material rise hundreds of kilometers above the frosted surface. They emanate from the tiger stripes, which are four in total, located in the south polar region and spaced about 35 km from one another. These are sub-parallel, linear depressions flanked on each side by low ridges. On average, each tiger stripe depression is 130 kilometers long, two kilometers wide, and 500 meters deep. The flanking ridges are on average 100 meters tall and two to four kilometers wide. The ridges that surround the tiger stripes are covered in coarse-grained, crystalline water ice, in contrast to the rest of the surface of the moon, which is covered by fine-grained ice. The ridges appear dark in Cassini's visible light images, extending several kilometers to each side. The very highest resolution images of the area show small features resting on rims and slopes of the undulating terrain. These objects appear to be boulders, but some may be more complex structures whose details are fine enough to be invisible at these resolutions. The Visual and Infrared Mapping Spectrometer (VIMS) instrument detected crystalline water ice along the tiger stripes. This creates an age constraint, because this ice would gradually lose its crystal structure after being cooled and subjected to the local magnetospheric effects. The change from crystalline to finer-grained, amorphous water ice is thought to take a few decades to a thousand years.

Our best tiger stripe images show phenomenal detail, considering the speeds and distances of the Cassini orbiter's encounters. However, they are not of sufficient high resolution to show whether ice towers exist. Cassini scientist John Spencer observes, "We certainly expect the water vapor to be freezing out at the vents. No one has talked about it building up towers and structures, but I don't think we can rule that out based on what we know and the images we have. Those Erebus towers would be just on the edge of resolutions that we have."

For a view of the terrain on the human scale, we must – for now – content ourselves with using analogs and informed imagination. When viewing Earth formations that may be analogous to sites on other worlds, researchers must be careful, as there are significant differences between conditions on Enceladus and Erebus. One obvious dissimilarity is air pressure. The air at the Earth's poles is thinner than that at lower latitudes. As mentioned before, the air pressure at the 3,800-meter Erebus summit is equivalent to a "fourteener," a 14,000-foot-high (roughly 4267 m) peak in Colorado. But the air pressure on Enceladus is essentially zero. How will cryovolcanic structures build and erode in the vacuum of near-Saturnian space? Some of the geyser vents appear to be linear, but others are focused and may be pinched off by structures more like conventional cones, or even like the Erebus towers.

The Southwest Research Institute's John Spencer has been studying many of the icy bodies in the outer Solar System. In light of his research on Enceladus, he cautions: "I think you will see structures around the vents due to vapor freezing to the surface. There will be fallback from the plumes, too. A lot of stuff comes out at high speed and that just escapes into space. Other stuff comes out more slowly and might gradually accumulate. But on Enceladus, around the vent itself, you see a trench, you don't see piles of stuff."

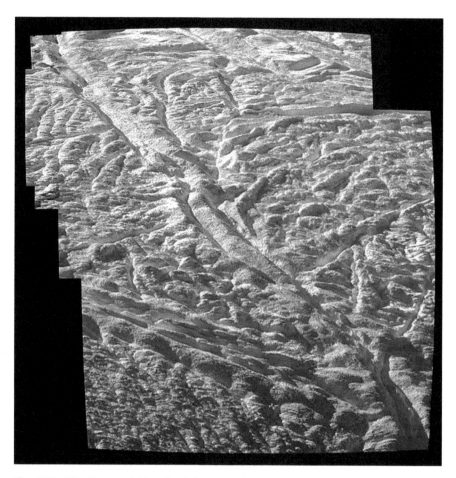

Fig. 6.17. The Saturn-orbiting Cassini spacecraft snapped this image mosaic – one of the most detailed views of its flight – of the volcanic provinces in Enceladus' southern hemisphere. Because of the spacecraft's high velocity and close range, smearing was digitally removed. This image was taken at an average range of 124 km (77 miles). Each pixel captures an area approximately 15 meters square (NASA/JPL/Cassini Imaging Team, composite by Astro0)

Yet another important difference is the erupting material itself. Volcanism is the process that brings magma to the surface but while on Earth, Io, and terrestrial planets the magma is molten rock, on Enceladus and other icy satellites the magma is mostly water. Spencer says, "I suspect we see rows of geysers as well as curtains of material. There are concentrations that produce more of a jet." These jets line up along the Enceladan tiger stripes.

The nuances of ice moon geology await further study, but the landscapes of the Harsh Continent, at least on a surface level, conjure visions of many exotic worlds distant to us in both space and time. From ancient Noachian Martian volcanoes to today's cryoeruptions on Enceladus, Erebus and its surroundings gift the keen observer with lessons about bodies beyond our planet. To really take advantage of those insights, we must dig into the details within the landscapes of other worlds.

Fig. 6.18. As our helicopter departs for McMurdo and leaves us behind on the flank of Erebus, we take stock of more volcanoes on more distant worlds. (Photo by the authors)

7

Landscapes on This and Other Worlds

Fig. 7.1. Ethereal towers of ice build above volcanic vents on the flanks of Antarctica's Mount Erebus, the world's southernmost active volcano. (Photo by Rosaly Lopes)

© Springer International Publishing AG, part of Springer Nature 2019
M. Carroll, R. Lopes, *Antarctica: Earth's Own Ice World*, Springer Praxis Books,
https://doi.org/10.1007/978-3-319-74624-1_7

Antarctica has been called the most otherworldly place on Earth. Its desolation reminds even the most seasoned of planetary experts of places like Mars or the ice moons of the outer Solar System. Its resemblance to these other worlds was the primary reason that we traveled to Mount Erebus: the volcano's Ice Tower Ridge and caves may well find counterparts on places like Enceladus and Europa, bitterly cold worlds ravaged by cryovolcanism, while its lava lake has many counterparts on Io.

Reasons for a Very Long Trip, Scientific and Artistic

A powerful tool in the study of other worlds is comparative planetology, the comparison of geological forms among planets and moons. When members of Ernest Shackleton's party summited Mount Erebus, they could have no idea that they were gazing upon landscapes similar to distant planets and moons, worlds completely unknown at the time. We now know better as scientists study terrestrial analogs on Earth that resemble the strange outer Solar System. Pressure ridges and fractures ripple across sea ice flows much like those found on Jupiter's moon Europa, and water has carved Norwegian fjords into the same kinds of shapes that methane rains have created along the shorelines of Saturn's moon Titan. The Earth's most beautiful glaciers find counterparts at the boundaries of Pluto's Sputnik Planitia, in ice floods on Saturn's Enceladus, or within the furrows of Neptune's Triton. Mars may have active rock glaciers (glaciers covered in gravel and rock debris) even today. The layering in Icelandic glaciers bears a strong resemblance to Martian polar caps, where seasonal dust storms leave layers of dark material each spring, and winter frosts lock them into a laminated record like the pages of a book. Volcanic flood plains in Africa find similar counterparts on Venus. The Channeled Scablands of North America's Columbia River find cousins on Mars as well, although they issue from different sources. In the Columbia flood, a glacial wall dammed the flow of a river for centuries or millennia. Eventually, the ice dam fractured and collapsed, releasing a series of floods. At their height, water flow exceeded ten times the flow of all the world's modern rivers combined.[1] On Mars, many such floods have been triggered by either the heat of meteor impacts (for example, Lowell crater) or the eruption of volcanoes, which melt subsurface ice (as in Kasei Valles, adjacent to the Tharsis volcanic province). Water on Mars may also have triggered or lubricated landslides. In her PhD thesis, Rosaly Lopes proposed that giant landslides around the Olympus Mons volcano on Mars were triggered by volcanic activity melting ice in the permafrost. Water has played a major role in sculpting the geologic features on Mars; however, the only place where water ice is exposed at present is in the north polar ice cap.

Antarctica's Erebus summit crater may look quite similar to ancient volcanoes on Mars, which were erupting at a time when water ice may have still covered surface areas. Erebus also provides insights into Saturn's geyser-ridden moon Enceladus, along with other volcanically active ice worlds (candidates include Saturn's Titan, Neptune's Triton, Uranus' Miranda and Ariel, and Pluto).

---

[1]Alt, David; Hundman, Donald W. (1995). *Northwest Exposures: A Geologic History of the Northwest*. Mountain Press

The Art of Science

Vast amounts of data from space exploration are returning to Earth. We have seen the huge Galilean moons of Jupiter from less than 200 km distance through the eyes of the Galileo spacecraft. The Cassini orbiter has finished a nearly two-decade reconnaissance of Saturn, its rings, and its moons. At this time, two Mars rovers, Curiosity and Opportunity, continue to explore the sands of the red planet Mars, revealing vistas of alien deserts, mountains, and craters. The Dawn spacecraft has now orbited asteroids Vesta and Ceres, and MESSENGER danced an orbital tango with the smallest planet, Mercury. The New Horizons has carried out exploration of Pluto on its way to a more distant Kuiper Belt Object, and ESA, India, China, and Japan have joined the club, with notable encounters of comets, Mars, Venus, and landings on the Moon, Titan and a comet.

The problem is that most data from spacecraft is not visual, but rather numerical. Streams of numbers come to scientists, who must then turn these numbers into something the rest of us can understand. That's where the artist and writer come in. When the Galileo craft charted sheets of invisible radiation and a vast torus of energy cocooning Jupiter, computer artists and painters showed us what these fields and particles might look like if they were visible to the naked eye, and how they relate to the king of planets. Information beamed back by the Mars rover Opportunity indicated that an ancient ocean washed across what is now rolling sand and rock. The artists rendered views of how this ancient Martian seascape might have appeared. Astronomical artists are also fleshing out information from exoplanetary systems, where our knowledge of extrasolar planets amounts to a few data points on a dimming light curve.

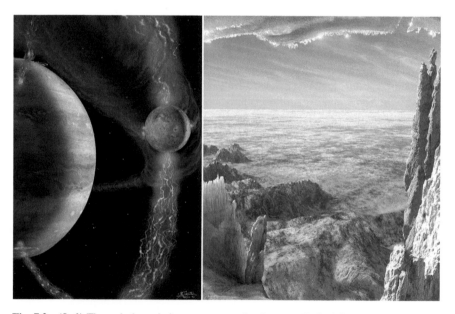

Fig. 7.2. (*Left*) Through the artist's eye, we see what is normally invisible: the energy fields linking Io to Jupiter through its flux tube. (*Right*) Artist's rendering of exoplanet Kepler 62f as a frozen ocean world. The available data consists of a periodic drop in a light curve and estimates of the planet's size, mass, and distance to its star. (Paintings by Michael Carroll)

The more data we have, the better we can guess at what is going on across these distant landscapes. Antarctica and arctic regions in the north provide insights into the moons and dwarf planets of the outer Solar System. Researchers have begun to focus on the large moons like Ganymede, Europa, Enceladus and others, as they are considered targets for astrobiological studies. But the Earth differs in significant ways from the ice moons. It has more gravity and – except in the case of Titan – more atmosphere. Jet Propulsion Laboratory Europa expert Robert Pappalardo cautions, "Beware of direct analogy." As an example, he points out that "sea ice is about a meter thick, and Europa's ice shell is tens of kilometers thick. On Europa, ice won't raft on top of itself as in sea ice."

Georgia Tech's Earth and Atmospheric Sciences Assistant Professor Britney Schmidt, whom we met in Chapter 5, agrees. She has been studying Antarctica and its surroundings as a testing ground for future exploration of ice moons like Europa and Enceladus. "It turns out that for Europa and some of the other ice moons," Schmidt says, "the strongest similarities are deep under the ice, under the ice shelf at what we call the grounding line. There is no perfect analog for what's going on in the ice moons, because most of the processes here are dominated by the silicates. Ice in Antarctica is produced by compaction of snow, which we don't have on Europa. There, we have stuff that's growing from below, so the bottom of the ice is the most important."

Schmidt and her colleagues study the underside of the ice shelf where the ice and water are interacting to see how the ice builds up and learn about the processes involved in its melting. She in suggests that sea ice is in fact a terrible analog for Europa. The reason that the sea ice is breaking up is that sea ice is very thin and is filled with brine channels. "It's not breaking up mechanically in the same way that the ice on Europa is," Schmidt says. "On Europa, the ice is at least three km thick, and probably more like thirty. The one or two meter thickness that sea ice is doesn't really begin to describe that. Just because it's shaped like stuff you've seen before doesn't mean it's the same."

Schmidt points to Europa's chaos regions, terrain that has fractured into plates and peaks of ice. Researchers have tracked ridge patterns on the surface, demonstrating that surface sections have separated, twisted, and rotated before refreezing into new arrangements. "What's happening in the chaos regions is a lot closer to ice shelf breakup. Those are pretty rare on Earth. Europa's case is really like iceberg calving, where thick ice is breaking because of fractures. On Earth, the interaction of fractures and a little bit of water is why that breaks up, but it's driven by the fact that it's melting, which is just not possible on Europa. It's a violent process, and so is iceberg calving: very violent, very energetic."

As Schmidt notes, ice shelf breakup is very rare. The Larsen C ice shelf on the Antarctica peninsula has been deteriorating for more than 20 years, and in July 2017, one of the largest icebergs ever recorded broke off the ice, attracting attention around the world.

In some ways, the pressure ridges (outside of Scott Base, for example) are similar on a smaller scale to what might be found on an ice moon because the sea ice is sheering. The frozen sheet breaks, pushes up, and piles atop itself. But we have no imagery with enough detail to see those kinds of features on Europa, Schmidt points out. "At those scales, it's not really clear what that's going to look like. We've seen Enceladus up close, and it looks really different from those. [Those chaos regions look a little like sea ice that's breaking up, but] if you really look at it and measure it, you can see that it's completely different. You don't see these huge cliffs in the sea ice, you don't see boulders like you do on Europa. So, you may see big patterns that look similar, but there's always something missing."

Schmidt and other researchers have been puzzled by the chaos terrain and other features that appear to have melted through from underneath. If Europa's crust is kilometers deep, how is the ocean breaking through all that solid ice? Schmidt hit upon the idea that Europa's ice crust is impregnated with subsurface lakes. If localized bodies of water reside near the surface, they could explain some of the features on the surface. "It solves some of the problems of a thick ice crust," Schmidt says. "It lets you break up the ice in specific places without needing a ton of thermal energy, and it explains a lot of the observations like the steep-sided cliffs." The Europa Clipper mission, due to launch in the early 2020s, will have a radar instrument capable of penetrating ice deep enough to search for these proposed subglacial lakes under the chaos regions.

Volcanoes Out There: The Mountains Close By

While drawing analogies between Antarctica and ice moons may be problematic, strong parallels do exist among the terrestrial geological forms and volcanoes on these distant worlds. To understand their erupting mountains, lava lakes, and fissures, we must see them in the context of volcanoes across the entire Solar System.

Volcanoes blossom across the faces of many planets and moons, and they have played an important role in sculpting their surfaces and atmospheres. Volcanoes enriched Earth's primordial atmosphere; much of the air we breathe today comes from the atmospheric raw materials of early eruptions. The other terrestrial planets Mercury, Venus, and Mars, along with the Moon, formed in much the same way as the Earth. While rocky volcanoes built the surfaces of the terrestrials, the development of the outer planets was ruled by primordial gases and ice. In the inner Solar System, these materials were cleared away by the adolescent Sun in an energetic epoch called the T-tauri phase. During this period, increased solar wind pushed light volatiles out into the distant regions of the planetary system, leaving only the heavy materials behind. Those materials formed the building-blocks of the rocky inner worlds.

All of the inner planets and the Earth's moon appear to have gone through at least early volcanism and magma oceans. Volcanic domes, cones, and flows have been found on the Earth's Moon and on Mercury. Venus and Mars display extensive volcanic structures created over long periods of time. Some of these structures may be dormant or active even today and are constructed the same way that Erebus is.

It turns out that Venus has her own versions of the Earth's Kilauea and Etna volcanoes. Among the terrestrial (Earth-like) worlds, Venus is *Volcano Central*. Venusian real estate is peppered with more volcanic shields, domes, cones, and flows per square kilometer than any other parcel in our Solar System. Most if not all are dormant or extinct, standing only as reminders of the planet's violent youth. Were it not for the gloomy, acid-laden ochre sky, many of Venus' volcanic landscapes would seem familiar to any Hawaiian. Other landforms are alien beyond the imaginings of the best science fiction writers. Flashfloods of molten rock have fanned out into smooth flatlands. Remnants of Venus' hellish childhood show themselves in long, sinuous channels that snake their way to the stone oceans. Some of them are thousands of kilometers long. The Baltis Vallis, in fact, may be the longest channel in the Solar System, over a kilometer wide at places and over 7,000 kilometers long. Both ends of the dried-up lava river have been obscured by newer flows; it may have been considerably longer. The plains of our cloud-covered neighbor play host to hundreds of thousands of small shield volcanoes. Some Venus shields have grown to titanic size, towering a kilometer or more into the sulfuric acid haze. Representative of these giants is Gula Mons. It rears up in the northwest region of Eistla Regio, a 6,000-km-long east-west trending highlands area that borders the great Aphrodite Terra continent. The mountain towers 5 kilometers above Eistla. Rugged scarps and troughs score its northern face, radiating out from the central summit caldera. The area may represent the most recent of lava flows, perhaps similar to lava flows on Earth (the rough, fragmented type of lava found on many volcanoes).

Other volcanoes known as ticks, anemones, arachnoids, and pancakes add to the alien world's Hadean ambience. Their bizarre nature may stem from other factors in the alien environment of Venus. Surface temperatures simmer at over 700°C (900°F). Air pressure at the surface is 90 times that of Earth at sea level – something more akin to the depths of Earth's oceans. Sulfuric acid drizzles in the lightning-laced clouds above. As real estate goes, it's not the prime landscape for a summer house.

Within this nightmarish realm lie the pancake domes: roughly disk-shaped, flat structures that range up to 15 kilometers in diameter but are mostly less than a kilometer in height. Their flat tops are often fractured. It is likely that the domes are the result of very thick lava flows that oozed out. Over 150 have been spotted on the planet. Pancake domes seem to lead to another form of volcano. The structures known as ticks appear to be degraded pancake domes, and they usually congregate along rift zones. Ridges radiate outward from the dome itself, often ending in landslides. The ridges terminate in sharp ends, giving the appearance of the legs of a tick. The legs may be scars of avalanches or the product of dikes running from the central body of the volcanic beast. The summits tend to be concave. Approximately 50 ticks have been located so far. Ticks bear some resemblance to terrestrial seamounts, undersea volcanoes whose flanks have degraded into radial ridges.

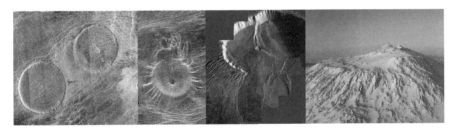

Fig. 7.3. A volcano menagerie: (*Left*) Pancake domes in Eistla Regio, Venus; (*Center Left*) a "tick" volcano; (*Center Right*) the grand Olympus Mons summit caldera; and the summit volcanic crater of Earth's Mount Erebus (All images courtesy NASA/JPL except Erebus: photo by Jack Green courtesy National Science Foundation)

Yet another member in the Venusian volcano parade is the anemone. Anemones exhibit overlapping lava flows arranged in flower-like petals radiating from the center. The lava patterns usually occur in association with fissure eruptions involving a series of elongated vents at the summit.

The arachnoids add their volcanic domes to the landscape in a different way: they are surrounded by a cobweb of fractures and crests. Russian scientists first named these features after seeing radar images of them taken by their Venera spacecraft, the first synthetic aperture radar systems to be orbited at Venus.[2] Vast, concentric fractures spread out from volcanic vents, ranging in size from 50 to 230 km. Lines radiating beyond the arachnoids may be cracks or ridges triggered by upwelling magma that stretched the surface around them.

The surface of Venus has been mapped from orbit using Synthetic Aperture Radar (SAR) on board the Magellan spacecraft. We cannot visually tell if there are active volcanoes belching plumes of smoke today. But radar imaging of Idunn Mons, a volcano in the area called Imdr Regio, has shown a volcanic construct at the site where infrared mapping from the European Venus Express mission revealed a possible "hot spot". The data may indicate recent lava flows. Venus Express results also showed that rocks from three volcanoes in regions known as Imdr Regio, Dione Regio, and Themis Regio were anomalously bright in the infrared compared to their surroundings, suggesting that they were relatively young and unweathered. Suzanne Smrekar from JPL, who led the analysis, concluded that flows in these areas were younger than 2.5 million years, and likely younger than 250,000 years – very new in geologic terms. Other evidence of volcanic activity comes from atmospheric data, which shows varying chemistry at high altitudes that is possibly indicative of volcanic outgassing on a large scale.

Venus has some titanic volcanoes, but the largest volcano yet discovered in the Solar System is on Mars. Olympus Mons, with a summit 24 kilometers high, is nearly three times as tall as Mt. Everest. The massive shield volcano's base would cover the country of Spain. Olympus Mons has a great deal of company. It stands on a colossal mound called the Tharsis Bulge, a dome built by multiple gigantic volcanoes. Tharsis rises 10 km above the Martian plains and spreads 4,000 km across. Dozens of volcanic formations dot the region, ranging in size from humble cinder cones to volcanoes in the class of Olympus Mons itself. On the opposite side of the planet, the volcanic Elysium region has its own respectable summits, and some of the youngest lava flows on Mars blanket its flanks.

Volcanoes Farther Afield

Beyond the terrestrial planets, exotic forces unfamiliar to Earth power the volcanoes, the most important of which is tidal heating. The Earth's Moon is a geologically dead world, a rocky ball pummeled by eons of meteors and comets. Researchers expected the small

---

[2]The first radar at Venus was carried out as a bi-static radar experiment by the Soviet Venera 9 orbiter (1975). More sophisticated radar rode aboard the US Pioneer Venus orbiter in 1978, leading to the first global map. The highest resolutions came later from the Venera 15 and 16, which had partial coverage in the northern hemisphere, and the US Magellan orbiter, which arrived in 1990.

Fig. 7.4. Artist's rendering of the pancake domes of Eistla Regio, seen from inside one of the faults transecting the region (Painting by Michael Carroll)

moons of the outer Solar System to be equally quiescent, but when Voyager 1 arrived at Io – a moon we too have visited throughout the course of this book — they were greeted by a surface torn asunder by volcanoes of every size and description. In place of impact craters and rolling dust plains, sulfur painted the little moon's face in golds, reds, and yellows. White frosts tinted freshly deposited black lava blankets, while yellow and orange sulfur compounds fanned out as tawny plains. Io's internal heat, left over from its initial creation and from radiogenic decay, was probably long faded, with much heat lost to the cold of space. As previously mentioned, it is tidal heating that powers the most volcanically active world in our Solar System. The gravity of Jupiter combines with the gravity of the other large Galilean satellites to force Io into a cosmic tug-of-war, causing the surface to rise and fall some 100 meters every day. This gravitational taffy pull creates heat on the inside of the little moon. Much of that heat comes out as volcanic eruptions, forming lava flows, lava lakes, and spectacular plumes. Io is constantly losing material to outer space and, while most of the material from the volcanic eruptions rains back to the ground in a deadly hail of frozen sulfur, about one ton escapes every second into the surrounding Jovian environment.

Io's plumes rocket up to 500 km into the airless sky. The volcanoes that sculpt the Ionian landscape are some of the most bizarre in our Solar System, but Io's lava lakes find a cousin at the center of Mt Erebus' summit (see Chapter 2). Scientists estimate that there may be over 500 active volcanoes across the moon's tortured face. This number is remarkable, considering that the surface area on Io is only equivalent to the North and South American continents combined. Thus, Io's small surface area may contain as many volcanoes as the entire land areas on the Earth, where there are currently about 500 or 600 active volcanoes (even more volcanoes on Earth rise from the ocean floor).

While the volcanoes of the inner Solar System and Io are forged in the furnaces of molten rock known as magma, they are not the only recipe for volcanic eruptions in the outer Solar System. Alien brews of frigid gases escape from some moons, while other eruptions are powered by "magmas" of exotic chemistry: super-chilled water mixed with ammonia, methanol, and other strange concoctions. These alien eruptions are behind the term cryovolcanism. This type of super-chilled eruption has left its mark on many places, from moons to dwarf planets. Next to Io, Jupiter's ocean moon Europa may have two types of volcanism for the price of one. Conventional volcanism may exist on its "sea floor" (at the junction of the silicate core and the ocean), perhaps similar to the submarine volcanoes in Earth's oceans. Tidal heating energizes the rock core of the moon, perhaps resulting in the flow of magma to the "surface" of the rocky center. But above that center lies an ocean perhaps tens of kilometers deep, kept warm and in liquid state by the same tidal force. The landscape betrays the sea's existence in many ways, including fractures and shifting ice blocks. Aside from visual clues, Europa generates a magnetic field consistent with liquid salt water. When the Galileo spacecraft flew within 346 kilometers of Europa's surface in 2000, its magnetometer charted a change in magnetic fields coming from Europa. These directional changes were very similar to those that would be produced by electrically conducting liquid within Europa's upper ice region. Unlike the electrical currents pouring from the Earth, Europa's field is induced—i.e. it is generated in response to Jupiter's mighty field lines. This induced field is continually changing in response to Jupiter's rapidly rotating magnetic field.

The existence of a vast European ocean is no longer in dispute, but the details are. Is Europa's crust a thin veneer, like a skin-deep apple peel a few kilometers thick covering the waters beneath? Or is it a substantial covering, essentially solid down to many tens of kilometers, perhaps perforated here and there by subsurface lakes? In either case, geyser-like plumes linking the interior ocean to the vacuum of space have been considered as a possibility. No such plumes were found in encounters by the Voyagers, the multi-year Galileo orbiter, or flybys of the Saturn-seeking Cassini spacecraft and the Pluto-bound New Horizons. But new data from the Hubble Space Telescope (HST) indicates that Europa could be exhibiting sporadic water eruptions through its ice crust. The Earth-orbiting observatory first glimpsed a nitrogen cloud hovering over Europa's southern hemisphere in 2012 erupting from near the south pole. Analysis suggested the plume was about 200 km (120 miles) high. Additional imaging obtained from HST in 2014 and again in 2016 revealed what appears to be a plume emanating from a 320-kilometer wide region near Pwyll crater, mapped earlier by the Galileo spacecraft's Photopolarimeter Radiometer as being warmer than its surroundings. In the latest data, the plume rises some 100 km above the surface of Europa. The discovery images imply that geyser-like activity on Europa is sporadic and unpredictable.

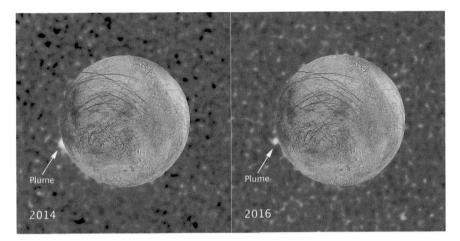

Fig. 7.5. HST composite images taken two years apart show a suspected plume of material erupting from Jupiter's moon Europa. The plumes appear to be sporadically active, erupting from the same region on the satellite. They are seen here in ultraviolet light, imaged in silhouette as the moon passed in front of Jupiter. The newly imaged plume (right) rises about 62 miles (100 kilometers) above Europa's surface. A Galileo image of Europa is superimposed on the plume data (NASA/ESA/W. Sparks (STScI)/USGS Astrogeology Science Center)

The Jewel in Saturn's Crown

Observers struggle to understand the details of Europa's proposed eruptions. In contrast, one moon that certainly displays spectacular geyser-like activity is Saturn's Enceladus. Enceladus is a bizarre little moon. Its surface is dominated by fresh, clean ice, giving Enceladus the most reflective surface of any world in the Solar System. This pristine ice surface is in places a tortured jumble of twisted ridges and cracked plains nearly devoid of craters. These plains appear to have been resurfaced, with some areas having a geological age of less than 200 million years. Ancient craters pockmark other parts of the surface. All surfaces of Enceladus—ancient or young—are very bright, suggesting that the entire moon is dusted with fresh material. Because it reflects so much sunlight, the surface of Enceladus only reaches a mean noon temperature of −198 °C (−324 °F), a little colder than other Saturnian satellites and certainly colder than Antarctica which at its coldest only gets down to −94.7°C (−135.8°F).

A scant 504 km across, the diameter of Enceladus would barely cover the Iberian Peninsula. Because of its diminutive size, Enceladus' geologically young surface was initially mystifying. By contrast, other nearby Saturnian moons had cold, dead surfaces, despite the fact that many are larger. The orbit of Enceladus is in orbital resonance with Dione, so Enceladus makes two orbits around Saturn for every one that Dione completes. From this, researchers assumed that tidal forces might be strong enough to trigger some kind of activity. Yet the problem was that Mimas, the moon next door, also has an irregular orbit, and it wears a geologically quiescent, ancient face with a giant impact crater that makes the little moon look like the *Star Wars* "Death Star." Some planetary geologists suggested that the interior of Enceladus might be heated by the rocking of the satellite, which in turn may be caused by the tug of nearby moons and the eccentricity of its orbit.

For decades, astronomers knew that Saturn's small, bright moon had some kind of relationship with Saturn's tenuous E-ring, but what that relationship was anyone's guess. Then, in 2005, the Cassini Saturn Orbiter snapped images of geyser-like jets gushing from the south polar region, coinciding with regions warmer than the surrounding terrains (Chapter 6). Cassini charted anomalies in the magnetic field marking the volcanoes' interaction with Saturn's magnetosphere, and stellar occultation observations locked down the density of the watery spouts and curtains. The jets, ejecting material at a velocity of about 60 meters per second, are massive enough to resupply Saturn's E ring.

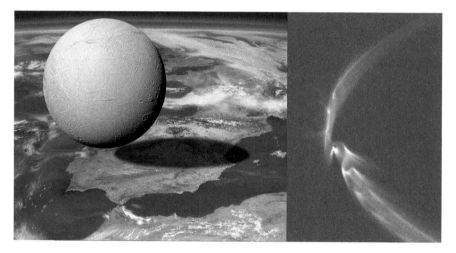

Fig. 7.6. (*Left*) Saturn's tiny ice moon Enceladus compared to the Iberian Peninsula (Art by M. Carroll); (*Right*) elegant filaments stream from Saturn's volcanically active moon Enceladus, bolstering material in Saturn's E ring (NASA/JPL/SSI)

The volcanism of Europa and Enceladus has far-reaching implications. It opens possibilities for finding active geology on bodies once thought to be too small for such processes. It also enlarges the number of sites for potential exobiology. Subsurface water on Enceladus may have been stable for long periods, perhaps millions or billions of years. Recent models and careful mapping have revealed that the crust of Enceladus is decoupled, a free-floating global shell completely separated from the rocky core by its hidden ocean. This fact, coupled with evidence for organic material at the eruptive sites, make Saturn's moon an intriguing target for astrobiologists. Tiny Enceladus thus joins the ranks of Mars and Europa as a place where future explorers may search for life among alien worlds.

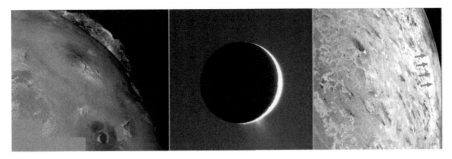

Fig. 7.7. Volcanoes compared: (*L* to *R*) Jupiter's Io spouts sulfur dioxide plumes, seen on the horizon; jets of water explode from Saturn's Enceladus; Neptune's Triton vents columns of nitrogen, a drifting trail marked by red arrows (Credits Io and Triton, NASA/JPL; Enceladus, NASA/JPL/SSI)

Titan: Volcanoes in a Methane Soup

The ringed planet hosts another moon that may feature active volcanism: Titan. The Cassini orbiter's results suggest that cryovolcanism has indeed been a significant geologic process on Titan and may be a major contributor to Titan's atmospheric methane. Several large flows spread across Titan's frigid landscape. Some of them are most likely cryovolcanic, but some appear to be fluvial, caused by liquid erosion from methane rainfall. Titan's surface shows plenty of branching channels, indicating that rivers of liquid methane do run there. The challenge is to identify which process caused a particular flow deposit. It is sometimes not easy to distinguish fluvial from volcanic in remote sensing data, and most of the Titan surface images, obtained by radar, are fairly low resolution (350 m at best) compared to the images we have from other worlds.

Several sites, changed over the course of Cassini's 13-year reconnaissance, and features indicative of volcanic structures have been found at sites like Tui Regio and Sotra patera. The most convincing evidence of a cryovolcano is at a site known as Doom Mons, which appears to be a volcanic mountain, its summit reaching 1.5 km above the surrounding plains. A circular pit resembling a volcanic caldera, known as Sotra patera, drops over a kilometer deep, itself nested within a roughly 500-m-deep indentation on the western flank. If Doom Mons is in fact a cryovolcano, it is the largest on Titan. Another possible volcanic mountain, Erebor Mons, is further north from Doom Mons, and the two are linked by flows interpreted as cryovolcanic. Mountains on Titan are named after those in Tolkien's *Lord of the Rings* books.

Appearances can be deceiving, but there are other lines of evidence suggesting that Titan may once have had cryovolcanic activity or may still have it. Foremost is the atmosphere. Titan's dense air is roughly 95% nitrogen, with a few percents methane. Methane is photo-dissociated in Titan's atmosphere; the Sun breaks the gas down so that it recombines with other constituents, forming organics like ethane, propane, and acetylene. This means that the methane is somehow being replenished. One theory proposes that Titan's large lakes of methane or ethane may be resupplying the atmospheric methane. At Titan's temperatures (a very chilly -176°C/ -285°F at the surface), methane behaves much like water on Earth. Several of the seas in Titan's northern provinces are the size of terrestrial bodies like the Black Sea, so substantial inventories of methane are exposed to the atmosphere. Titan's methane precipitates out of the sky as rains and mist, collecting into surface lakes. Some of it undoubtedly makes its way to the interior, where it could gather into reservoirs. It may also be percolating up from the liquid ocean in the interior of the moon. Methane from these sources could escape and contribute to the atmosphere. However, these sources may not supply enough: a more likely replenisher of the methane is cryovolcanism.

Scientists expect any cryovolcanic features on Titan to be somewhat different from those on other satellites such as Triton and Europa, because Titan's thick atmosphere plays a significant role. The atmosphere has three effects: first, conditions make it harder for gases to come out of the cryomagma. Second, it affects how far away from the vent any explosive products land (projectiles cannot travel as far as in a low-gravity vacuum). Finally, the dense atmosphere would cool cryolavas faster than those in an environment devoid of atmosphere. Some researchers predicted that Cassini would discover domes similar to the "pancake" domes on Venus, but unless the features are very eroded, these have not been found so far. Researchers also anticipated that cryomagmas would likely be a mixture of water ice and ammonia, perhaps with a methanol "chaser." Laboratory experiments suggested that these cryomagmas would be quite viscous, perhaps the consistency of wet concrete. Some of Titan's flows do appear to be of high viscosity.

Sotra patera, the pit adjacent to Doom Mons, is considered the best candidate for a cryovolcanic crater on Titan. The area has more going on to raise researchers' suspicions. The Sotra patera/Doom Mons region showed changes in brightness from one Cassini pass to another in the course of one year. Changes in the surface brightness were noted by Greek researcher Anezina Solomonidou using Cassini's Visual and Infrared Mapping Spectrometer (VIMS). The surface variations, along with the flow-like formations associated with Sotra, suggest that the region may currently be volcanically active. The regions that have changed appear to be chemically different from the surrounding landscape, bolstering the idea of material erupting onto the surface.

VIMS data is consistent with deposits of ammonia frost, which would be transient, evaporating away fairly quickly. Models indicate that ammonia is a major component of Titan's interior makeup, and its presence in the changeable sites further implies that the Sotra Patera region may be venting material from the interior of Saturn's largest moon.

The European Space Agency's Huygens probe, which landed on Titan on January 14, 2005, obtained other clues that cryovolcanism may be occurring on Titan. Although the amazing surface images did not show any features that were obviously cryovolcanic, the probe obtained a surprising finding after its landing on the surface. Huygens revealed a specific type of Argon (40Ar) in Titan's atmosphere. This type of gas shows that Titan is bringing material to the surface from deep inside of itself. Cryovolcanism would be one means by which this material might be brought up.

Fig. 7.8. Artist's rendering of a proposed cryovolcanic region, Hotei Arcus, on Titan. Ammonia-brightened viscous lava flows leave soft ridges in the foreground, while methane drizzles from the sky in the distance (Art by Michael Carroll for JPL)

Cousins in the Darkness: Triton and Pluto

Beyond the orbit of Saturn, out at the edge of our planetary system, frigid temperatures lead to a unique interplay of geology and atmosphere. Even in the cryogenic temperatures of the Solar System's outskirts, eruptions take place. Active volcanism plays a key role in the environment of Neptune's largest moon, Triton.

Triton was Voyager 2's last agenda item before it went on its way to the outer edge of the Sun's heliopause and into interstellar space. The intrepid craft flew above Triton's south pole on August 25, 1989, taking stereo images that showed two dark, tall plumes, reaching about 8 km above the surface and leaving trails for about 150 km. Other images of Triton's southern polar region revealed more than 100 dark, streaky deposits, presumably a result of other plumes, implying that plume activity must be fairly common. We don't know how widespread this strange volcanic activity is on Triton, as the northern polar provinces were in darkness during the Voyager flyby. At lower latitudes, images showed a peculiar terrain with flow-like features, probably resulting from cryovolcanic flooding of older topography. This terrain has been dubbed "cantaloupe" because of its resemblance to the skin of the melon.

Triton is a curious satellite. Its orbit is inclined to Neptune's equator and travels in a retrograde path (opposite to Neptune's spin direction). Since nearly all moons circle their parent planets in the same direction as the planet's spin, any moon that orbits in a different direction is suspected of being a captured object that formed somewhere else. Triton's "backwards" orbit suggests that it did not form near Neptune but is indeed a captured object. The moon is very, very cold – only -235°C/ -396°F at the surface. This is well below the freezing point of nitrogen, the material that makes up the south polar cap. Both nitrogen and methane have been spectrally identified on Triton's surface, and carbon monoxide and carbon dioxide have been seen in the mix, too. Water ice is thought to make up the crust, but it has not yet been specifically detected. Triton's thin atmosphere consists largely of nitrogen. The atmosphere transports nitrogen ice from pole to pole every Triton year, keeping the surface temperature nearly the same everywhere.

Triton's southern hemisphere was just approaching summer solstice when Voyager flew by, an event that occurs only once each 165 Earth years. Triton's orbit is not round, but oblong. Because of this, the path it takes around Neptune wobbles around the planet, much as a top rocks back and forth as it spins. This orbital rocking is called precession. Triton's orbit precesses every 688 Earth years, causing the latitude of the warmest spot on the moon to change every year. The region receiving the most sunlight on Triton wanders as far as 55 degrees from the equator. Voyager happened to fly by in one of those extreme years, when the Sun was beating down on the rarely illuminated polar wastes. The active plumes were located close to the area where the sun was most directly overhead.

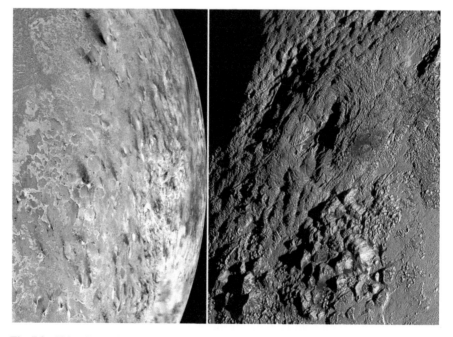

Fig. 7.9. Volcanic provinces compared: (*Left*) At Neptune, Triton's geysers release dark plumes that drape blankets of material across pink nitrogen ices; (*Right*) Pluto's Wright Mons rises in concentric flows—perhaps cryolavas—above the plains of Sputnik (near center of image). Reddish material may be organics precipitated from the interaction of atmospheric methane and sunlight. (Triton image courtesy NASA/JPL. Pluto image courtesy NASA/JHUAPL/SwRI)

Unlike the volcanoes on the moons of Jupiter and Saturn, which are ruled by tidal heating, Triton's plumes may be powered by sunlight. Most of the volcanoes that Voyager 2 spied on Triton were located near the southern pole. This area receives the majority of solar radiation, and it appears that the modest amount of heat generated on Triton from the distant sun may play a key role in the volcanic activity on the moon. Like Enceladus, Triton's plumes are similar to geysers – more Yellowstone than Vesuvius – and may be caused by solar sublimation (melting directly from ice to vapor) of the polar cap. Pink nitrogen ice on the surface cocoons pockets of nitrogen gas. Nitrogen ice is very clear, so sunlight can penetrate deep into the ice. Solar energy is absorbed and trapped by dark, carbon-rich impurities a few meters below the surface. This mild heating is enough to cause the interior of the nitrogen ice to become gas, as even a small increase of about 10° C over the surface temperature is enough to raise the vapor pressure of nitrogen by a factor of 100. The warming gas expands and explodes into the near-vacuum of Triton's environment. If this model is correct, cryovolcanism on Triton is a side-effect of sunlight rather than an internally driven phenomenon. Obviously, this process has a terrestrial analogue: the greenhouse effect. The greenhouse effect is not limited to particle-clogged atmospheres. It can sometimes operate within a solid, especially when that solid is nitrogen ice. Hence, the process that drives volcanism on Triton is called the "solid state greenhouse effect." Triton also has tectonic features such as ridges and valleys. There are very few impact craters. This indicates that the moon's geology is dominated by internal processes and that the surface is very young.

Fig. 7.10. Artist's rendering of the strangely calm nitrogen columns issuing from Neptune's moon Triton. Formations are based on the Erebus towers (Art by Michael Carroll)

Triton's activity is so different from the activity we see on other planets and moons that some researchers prefer not to classify it as cryovolcanism. But other areas on its surface do display features similar to more classic volcanism. Some circular plains are surrounded by cliffs, which stair-step down to a smooth floor. Some of these plains have been flooded, perhaps by cryovolcanic flows. One in particular, Ruach Planitia, appears to have frozen waves across its flat base. At its center lies a collapsed pit that shares characteristics with volcanic calderas.

From the clues of its retrograde orbit and its bizarre chemistry, scientists consider it likely that Triton came from the Kuiper Belt, a band of comets and rubble that ranges from outside the orbit of Neptune to a distance of over 4 billion miles from the Sun. In the years after the Voyager encounter, it was thought that Triton might be a similar world to the dwarf planet Pluto, poster child of the Kuiper Belt. But the New Horizons encounter of the Pluto/Charon system demonstrated that Pluto has a truly distinct nature. The highly antici-pated results of the New Horizons flyby of a more distant Kuiper Belt object in 2019 may shed some light into these far-flung, frigid worlds.

Spanning less than a fifth the diameter of Earth, Pluto was widely expected to be a cra-tered, geologically dead world. But far from a quiescent ball of ice, Pluto sports a host of dramatic and puzzling features. Some Plutonian ice mountains rise to heights comparable to the Rocky Mountain range. Pluto's peaks are dusted in ices of nitrogen and methane, while frozen gases become ice floes that wander from mountain valleys, spilling into vast ice plains. The largest ice plain region, the heart-shaped Tombaugh Regio, is named after Pluto's discoverer, Clyde Tombaugh. The western side of Tombaugh is frosted in carbon monoxide ice, while nitrogen and methane ices encrust other areas. The central region of the "heart" is a high plain called Sputnik Planitia. The province has fractured and eroded into polygons, called cells. To the south, the cellular plains become pockmarked by pitted terrain. The hol-lows may be related to sublimation of ices. Dark streaks run in parallel, perhaps the product of fierce winds in Pluto's thin atmosphere or a process that bears some resemblance to Triton's solid-state geysers. Methane ices erode tectonic faults, cliffs, and crater rims.

Just east of Tombaugh region lie some of the strangest formations in the Solar System. Here, "bladed" features slice through the hills of Tartarus Dorsa. They rise like scythes from the hummocky terrain, running as parallel fins aligned north to south. The blades tower hundreds of feet high, with gaps of a few miles in between.

Pluto's nitrogen/methane atmosphere ebbs and flows as the planet follows its eccentric orbit, dipping just inside the orbit of Neptune (at 29.7 astronomical units, or 29.7 times Earth's distance from the Sun), and then circling out as far as 49.3 AU. This shifting dis-tance means that as the planet nears the Sun, its ices sublimate. Complex hazes drift within the rarified air above the ice world. Methane reacts with solar radiation to create organic compounds. These reddish "tholins" fall from the sky, blanketing the ground with ruddy rivers and stains.

Pluto appears to have at least two major volcanic structures. Informally named Wright Mons and Piccard Mons, the mountains exhibit a series of concentric ridges and fractures surrounding what appear to be summit calderas—volcanic craters—at the top. They are large as planetary structures go and the largest candidate cryovolcanoes so far found in the Solar System: Wright Mons rises some 4 kilometers above Pluto's Sputnik Planum, while its base spreads to a diameter of roughly 150 km. The depression at its summit has been compared to the calderas of Mauna Loa on Earth and to Mars' Olympus and Arsia Mons, all of which are shield volcanoes. Only one impact crater has been charted on the mountain's surface, implying that the formation is quite new and may even be active today. A similar structure, Piccard Mons, lies to the southwest. Although it was in twilight during the encounter, image enhancement shows a shield-like mountain with summit depression similar to that seen more clearly on Wright Mons, and the two are very similar in size. Perhaps these twin peaks mark Pluto's version of the Tharsis Bulge. Pluto's cryovolcanoes most likely erupted a slurry of substances such as water ice, nitrogen, ammonia, or methane. However, like the identification of cryovolcanoes on Titan, the evidence is not totally conclusive, and it is likely that we will have to wait for future missions to tell us for sure.

More Than Rocks

Airless Europa and Enceladus will be sporting no fat-insulated Weddell seals on their surfaces. But beneath the ice, it's anyone's guess, and Antarctic biology gives us ideas about what visages life might take in the ocean of an ice moon. One species of anemone, *Edwardsiella andrillae*, embeds its body within solid ice, letting its tentacles drift in the frigid waters below the sea ice. Blue blood courses through the veins of polar octopi (some 16 species live in Antarctic waters). Rather than the rusty iron that tints human blood red (hemoglobin), cephalopods like octopi use blood based on copper. Their blue bodily fluid is stained by a pigment called hemocyanin, which enables the tentacled creatures to thrive in the near-freezing Antarctic seas.

Fig. 7.11. In the Crary Lab, the world's largest single-celled organism, visible to the naked eye, floats in a vial. The creature is a perfect example of polar gigantism (Photo by the authors)

They grow 'em large in Antarctica. Here, sea spiders – humble in size in warmer waters – grow to the size of a small dinner plate. The phenomenon is called polar gigantism. As early as a century ago, zoologist C. F. Laserson reported odd cases of gigantism during the Australasian Expedition (1911-1914). "…we saw [the dredge] full and overflowing with every form of sea life it is possible to imagine." Laserson singled out the red sea spiders that were "from six to seven inches across the legs."[3] Polar gigantism is still a biological mystery. Many theories have been put forth, including the liberation from physical and environmental constraints, systematic changes in biological development, and species-wide effects from various ecological and evolutionary processes. It crops up in many forms. Insect-like isopods, typically smaller than a sesame seed, grow to 5-inch-long behemoths in the polar cold. The arms of sea stars may span more than a foot across. Scallops living within the sea ice grow larger than their cousins in more tropical climes, but it takes them a while. Growth in extremely cold conditions is slowed – a three-inch scallop may be 20 or 30 years old. But the variety of creatures thriving beneath and within the sea ices of Antarctica demonstrates the tenacity of life, and the possibility of its existence even in the cryogenic realms of the outer Solar System.

The biology of Antarctica informs our educated guesses and theories about the new frontier beyond our planet. The sciences of geology and geophysics go hand-in-hand with the life sciences, providing insights into worlds across the Solar System. From the tiger stripes of Enceladus to the bladed terrain of Pluto, from the pancake volcanoes of Venus to the summits of Mars' Tharsis province, the volcanoes, glacial flows, mysterious caves, and rugged mountains all find analogs in Antarctica. Such vistas await the courageous explorers of future generations.

---

[3] *South with Mawson*, by C. F. Laserson, Angus & Robertson 1957

8

Future Explorations

Fig. 8.1. Explorers will leave their mark on many landscapes across the Solar System (Photo by authors)

© Springer International Publishing AG, part of Springer Nature 2019
M. Carroll, R. Lopes, *Antarctica: Earth's Own Ice World*, Springer Praxis Books,
https://doi.org/10.1007/978-3-319-74624-1_8

Alfred North Whitehead said, "Without adventure, civilization is in full decay." Certainly, Whitehead's notions align with those of Amundsen, Scott, and other explorers of the heroic age. A new generation of pioneers shares them as well, whether they explore the deep frontiers of the Earth's oceans, the polar wastelands of the arctic and Antarctica, or the frontiers beyond the Earth's atmosphere. Even today, Antarctica displays the essence of adventure, exploration, and research in the wider context of the cosmos around us. McMurdo shadows what may be in our human beachheads on Mars, while the sea ice pressure ridges, glaciers, and strange volcanic towers in the Antarctic wilderness may well presage travelers' vistas as the human race makes its way further afield.[1] Just what will those explorations look like? From the sands of Mars to the frozen nitrogen wastelands of Pluto, engineers and futurists are filling in the details as our technology matures toward that end.

Mars

For its part, Mars intrigued us at the outset. Its Earthlike seasonal tilt is close to that of the Earth's, as is its length of day (just 37 minutes longer). Through the telescope, we can watch its polar caps – the Martian version of the arctic and Antarctica – as they expand in its winters and shrink with the onset of its summers. The mysterious, undulating dark regions that expand with the shrinking ices caused many early astronomers to assume that a wave of vegetation was flowering in the Martian spring.

Flagstaff astronomer Percival Lowell propagated the idea of there existing Martian canals, born at the hands of an intelligent alien race. He crafted detailed maps of Martian canal networks and "oases," and wrote popular speculative articles and books about life on Mars. Authors like H. G. Wells, Edgar Rice Burroughs, and Ray Bradbury took up Lowell's torch, nurturing Western culture's infatuation with the red planet. Most of that Mars – the 19th century desert world covered by dwindling canals and dying civilizations – has faded before the eyes of our spacecraft and advanced telescopes. The flowering of Martian jungles has alas turned out to be the shifting of volcanic dust as it tints wide swaths of Martian territory, moved by seasonal winds. But our desire to find life beyond Earth, even in microbial form, continues to inform our priorities for space exploration and to fire our dreams. Our most recent images from surface rovers and super-orbiters reveal landscapes reminiscent of sub-Saharan Africa or America's desert southwest. It's a place with Earthlike qualities, and we want to go.

Many aerospace strategists see Mars as a logical steppingstone along the way to farther shores. Again, Antarctica continues to play a role in the exploration landscape. Its seas and ices train engineers and explorers for challenges on other worlds. The elements of severe temperature, relentless winds, extremes in the day/night cycles, and operating equipment in hostile environments all help us to hone our skills for those alien environments.

---

[1] For more on human exploration, see *Living Among Giants: Exploring and Settling the Outer Solar System* by Michael Carroll

Lessons learned here are already making their way into off-world outpost engineering studies (see Chapter 4 for the Mars component of exploration and settlement).

Mars serves another parallel with its rich and varied volcanic history. Vast shield volcanoes rise from its plains. Cinder cones pepper its landscapes. We have even found evidence of hydrothermal interactions where water has come into contact with magma, conjuring up comparable images of Earth's Yellowstone and Erebus.

Settling a Higher Frontier

The icy moons and dwarf planets may seem inhospitable places for life, whether you are a human or a microbe. The outer Solar System is dark and viciously cold. Distances make communication and travel difficult. While an average message gets from Mars to Earth in less than 15 minutes (radio waves travel at the speed of light), the same communiqué would take about 80 minutes if it were to come from the moons of Saturn.[2] But the outer worlds, giants of gas and ice, possess entourages of icy and rocky moons replete with water, minerals, and hydrocarbons. In addition to their abundant natural resources, they hide deep mysteries that will likely transform our foundational understanding of our planetary system. Carolyn Porco, Principal Investigator for the Cassini Saturn Orbiter imaging team, says: "No matter how you measure it, whether you count the number of bodies or the volume taken up by their orbits, the vast majority of our Solar System lies out beyond the orbit of the asteroids." These worlds in all their wild variety display the wonders of a new frontier worth exploring.

At Jupiter, king of the worlds, moons with a wide spectrum of conditions will beckon future travelers. Mount Erebus' glowing lava lake reminds us of Jupiter's innermost Galilean satellite Io. Instead of palm trees and plains of wheat, the little world has been torn asunder by volatile geyser-like spouts and exploding volcanoes. Its ferocious erupting sources have painted the moon's face in ochers, yellows, and reds. On the scale of human experience, Io's tawny landscapes must share the grandeur of the Russian steppes or America's wheat-covered plains: undulating golden vistas peppered with soft greens and punctuated here and there by the black and orange of sulfurous pits. Io's snows come at all times of the year; waterless blankets of sulfur dioxide frosts clinging to its rocky ground. Most of the surface's brilliant hues can be attributed to sulfur and sulfur dioxide. Laboratory simulations show that sulfur transitions through dramatic color changes as it cools. Molten material is most often black, changing as it cools into red, orange, and yellow. Plume deposits are seen to change color with time, with reds fading to yellows. Sulfur dioxide frost powders some of the landscape in snowy white and pale yellows. Inside some of the volcanic craters, bright greens are seen, nicknamed by researcher Paul Geissler as "Io's golf courses." The greens are likely due to the interaction of sulfur compounds and hot lava, or they could be due to lavas rich in olivine, the mineral that is found in abundance at Hawaii's Green Sand beach.

---

[2] These are average times, as the distance between planets changes as they circle around the Sun, each with a different speed.

Fig. 8.2. The immense molten lava lake at Loki Patera, Io, seen from an imaginary overlook. Eruptions issue from one end, while nearby moons Europa and Ganymede float in the sky of this artist's conception. (Art by Michael Carroll, courtesy collection of Jani Radebaugh)

Finding a safe harbor to put down roots on hellish little Io is another matter. Io's leading hemisphere is partially sheltered from radiation, as Jupiter's plasma comes up from behind it, overtaking it on each orbit. Still, surface radiation at any location will be deadly within hours. Shelter might be found underground within some of Io's lava tubes and caverns, but the danger of returning lava might be substantial. Still, the view from Io would be incredible, with Jupiter's ever-changing clouds marching across its globe – spanning as far across as 39 full moons in Earth's sky – and the constant interplay of the other moons.

Exploring the Ice Moons

Alongside Io circle the icy Galilean moons of Jupiter. The icy moons of the planets even farther out offer vast opportunities for human exploration. With the advent of sophisticated missions like the Jupiter-orbiting Galileo and Saturn's Cassini, planetary scientists are beginning to understand the icy satellites as a cohesive group, a family of related bodies. The discovery of volcanism on Enceladus reinforced what planetary scientists knew might be possible. They had seen activity on Io and noticed hints that there might be activity on Europa and Enceladus from the early Voyager data, along with further data from the Hubble and other telescopes. Spacecraft and modern remote sensing have brought a sense of pattern to the variety of small icy worlds, but within those trends, puzzles continue to challenge observers. Our modern view of these varied moons now includes subsurface oceans, spectacular flowing ices, and many sites of current or ancient cryovolcanism.

For human visitation, Europa is the closest. The frozen world constitutes an easier target for future human exploration than Io. Low radiation zones exist above and below the equator on the Jupiter-facing part of the leading hemisphere. Its bright cliffs and ridges would be seen meandering to the horizon. In the south, plumes puff water vapor into the sky, although we still don't understand their schedule or trigger. Like Io, Europa would have a spectacular view of Jupiter, and its astrobiological possibilities will certainly keep it at the top of future explorers' bucket lists.

Jupiter's other ice moons, Ganymede and Callisto, are considerably larger, with Ganymede's diameter measuring greater than that of the planet Mercury. Lower radiation levels than at Io or Europa, due to their distance from Jupiter, may make these the first of the Galileans to be visited by astronauts. Callisto's cratered face is ancient and tattered. Ganymede's is a hybrid between Callisto and Europa. Old, heavily cratered terrain often ends in steep cliffs. Beyond this dark surface spread line upon line of icy ridges and cliffs, many resembling the ribbed landscapes of Europa. Parallel valleys and ridges, folded trackways of ice, plow across the face of Ganymede for hundreds of kilometers, slicing swaths through the darker terrain, breaking it into gigantic polygons. The ridges themselves are not made like many mountain chains on Earth, where plates shove together and force wrinkled ground upward. Rather, these valleys form as slabs of ice break apart along fault lines and lean over like books on a shelf. Many craters have been split, pulled apart, and otherwise obliterated by the shifting of great "plates" of surface ice.

Callisto's battered face is geologically the oldest of the Galileans, with ancient craters probably dating back to the initial heavy bombardment, a rain of meteors and asteroids during the formative years of the Solar System some 3.8 to 4.6 billion years ago. Its pervasive craters indicate that Callisto has been geologically dead for nearly as long. Callisto has no traces of Io's volcanoes, no hint of Europa's leaky plumes or ridged surface, none of Ganymede's bright calligraphy of troughs and crests. Callisto is a world unto itself. Monument Valley-like buttes rise from its icy mahogany plains, perhaps casting long shadows from pinnacles of pure water ice. Spectral tones may play across the landscape within those shadows, painting the moon's ancient, soft craters. Its bleak beauty and scientific importance in the Galilean family will draw researchers and tourists alike to its distant shores.

Our trip up Mount Erebus continually brought to mind Enceladus with its majestic plumes and frozen wilderness. Saturn's active moon offers astrobiologists free access to what's beneath the crust. No drills or submarines are needed: just a sampling device and a flight through the 400-km-high jets of seawater. Perhaps the greatest obstacle to exploring Enceladus is that of contamination. As with traffic from McMurdo to the Ross Sea Ice shelf, any probe that lands there will need to "shake the dust off," undergoing careful sterilization to avoid introducing any terrestrial and potentially invasive species to the moon. This decontamination process will also ensure that any signs of life subsequently found by our probes are not in fact ones carried from Earth but rather are indigenous Enceladus organic material.

Once human explorers reach that far shore, they'll witness jets and curtains of material soaring hundreds of kilometers above their heads. As these plumes rise from the base of the tiger stripes (see Chapter 7), they may even build towers akin to those on Erebus. Rather than the cratered landscapes common to many other moons, Enceladus is blanketed in frost, boulders, and hummocky valleys. Saturn and its rings will stretch across a third of the sky. Enceladus has something else going for it: virtually no background radiation. Aside from a hard vacuum and cryogenic temperatures, the moon enjoys a fairly benign environment.

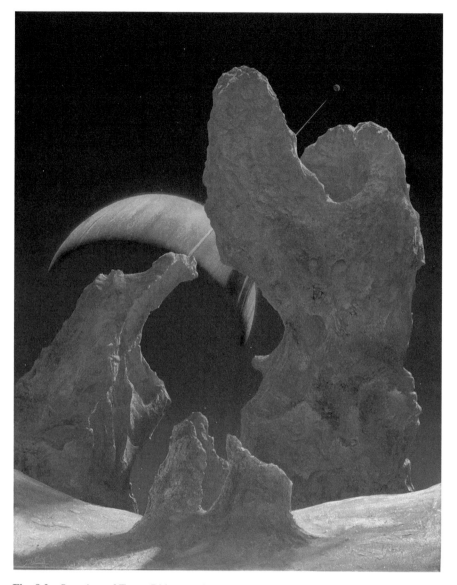

Fig. 8.3. Our view of Tower Ridge continually brought to mind icy moons like Enceladus. Here, the artist has based forms on those surrounding the Erebus plumes. On Enceladus today, jets erupt from the southern polar regions, but some terrain farther north shows signs that ancient activity may have occurred. Those ancient sites may have migrated north as the free-floating crust of Enceladus shifted, as in this scene. (Art by Michael Carroll)

What vistas face future astronauts on the other small ice worlds? The surfaces of these moons may well be reminiscent of Apollo landing sites. Saturn's Mimas, Tethys, Dione, and Rhea will look quite familiar from close up, as the ice at those temperatures behaves like rock. Craters will dominate the landscapes on a small scale, except for the occasional giant cliff. On the trailing sides of Tethys, Rhea, and Dione, a layer of dirtier ice covers everything except for the steep slopes and ice cliffs.

The visual similarity between icy satellites and the Earth's Moon results from similar cosmic erosion. The primary erosion on the Earth's moon today is from a constant drizzle of micrometeorites, which wear down the surfaces over long periods. Apart from this process, there is another Saturnian moon that has erosion of a different kind: wind and methane rain.

Fog-enshrouded Titan has its own highlands and mountains, canyons and valleys, lakes, rivers, and seas. The large seas on Titan have rugged coastlines analogous to the fjords of Scandinavia. A few of the largest lakes also have some irregular shorelines, but the smaller ponds are of a very different type. Many have circular or curving shorelines with steep margins. Because of the wall-like borders of these features, some researchers suggest that these lakes are the result of collapse or melting, much like the rounded lakes caused by melting ice blocks left behind by retreating glaciers on Earth. Geologists call this type of terrain *karstic*. On Earth, similar regions are very porous and often fractured, with groundwater flowing beneath their surfaces. On Titan, these lake regions may drain into subsurface methane aquifers that make their way to the lower elevations, eventually feeding the seas. That methane may also play a part in cryovolcanism. As we have seen, researchers have identified several candidate sites. And while the vacuum of Enceladus may preclude ice towers, Titan's environment certainly does not.

Titan's bizarre environment is actually more Earthlike than Mars or Venus is. The atmospheric pressure is half again as high as Earth's at sea level, so astronauts will not need a pressure suit, just good insulation and an oxygen supply. Exiting a spacecraft on Titan, an explorer will wear an environment suit far less elaborate than the Apollo or ISS suits. On a very clear day, with the right polarizing visor, one might just make out the shape of Saturn through the haze.

Saturn's mid-sized satellites find company out at Uranus. The two most geologically complex of the five major Uranian moons, Ariel and Miranda, have endured violent pasts. Both are cratered, but Ariel's pummeled plains have been scored by criss-crossing valleys and floes of material. Some bright areas may even be the remnants of cryovolcanic activity. Its sibling moon Miranda, smallest of the five, displays a surface rumpled and wrinkled by oblong "racetrack" regions (called coronae) and dramatic chevrons of uplifted terrain. Miranda also has the distinction of sporting the highest cliff known in the Solar System. Verona Rupes towers 5 km above Miranda's hilly, cratered plains. For extreme sports like BASE jumping or rock climbing, Miranda will be top of the list. Its awesome vistas, spectacular views of Uranus, sheer topography, and exotic geology will keep researchers and adventurers busy for centuries to come.

Now we return to Triton, the largest moon circling Neptune. Across that tantalizing hemisphere, observers have identified only 15 impact craters. This lack of craters underlines just how active and geologically young the surface of Triton is. Much of the moon's territory is blanketed by peculiar pocked flatlands called cantaloupe terrain. The cantaloupe terrain is unique in the Solar System. Stacks of plate-like forms are reminiscent of saucers arranged on a table with their upturned rims. One of the last high resolution

Fig. 8.4. What scenes will future travelers witness on Saturn's moon Titan? (Art by Michael Carroll)

pictures that Voyager snapped captured an oblique view across the cantaloupe terrain. It provides some idea of what that landscape will look like to future Shackletons and Scotts. Ridges cut through the plates, but the ridges are all warped, as if a crazed welder assaulted them with a blowtorch. Triton will be a fabulous place to explore. From the sites of the plumes, views of Neptune will be breathtaking, even to the most seasoned of travelers. Because of Triton's inclined path around the planet, the thin rings of Neptune will stand out against the ebony sky. The rings are in constant change, evolving and morphing over short periods of time. Unlike the ring systems of the other giant planets, Neptune's are clumpy, adding to the striking view. There may be active cryovolcanoes on Triton's surface, forms different from the geyser-like jets. Triton's cliffs are fairly low, and the moon appears to have no high mountains, but it will be a marvelous place to wander. Wispy layers of haze betray the presence of an atmosphere, especially toward the horizon.

As previously mentioned, Triton was long thought to be a twin of Pluto. Both bodies reside in the Kuiper Belt region of our Solar System, and both are about the same size. But two planetoids could not be more different.

Pluto: Ice Dwarf on the Edge

When New Horizons sailed by Pluto in July of 2015, its array of instruments unveiled a complex world of ice mountains, glacial flows, cratered realms and vast arctic wastelands. Pluto's extraordinary "heart", the Sputnik Planitia, provides researchers with clues to a hidden ocean there. Pluto and its large moon Charon face each other in a tidally-locked dance. Sputnik's vast plain lies on the hemisphere directly opposite of Charon. Its location tells planetary modelers that Sputnik is heavier than its surroundings. If Pluto has a buried ocean, its water would be denser than the icy crust above. Researchers estimate that this ocean may be extensive, perhaps even global, and up to 100 km deep. Its waters may be saltier than the dead sea. Other ocean models assert that beneath Pluto's Sputnik lies a sea of icy slush. Whatever its composition, Pluto's hadean seas will draw future explorers to their subsurface shores with drilling cores and submarine probes.

The weird world has even more to offer. Great ice mountains float upon polished plains of ice, and at least two structures may be the product of cryovolcanism (see Chapter 7). Long thought to be nearly airless, Pluto's skies display a host of atmospheric phenomena, including complex haze layers and ground fog. For a small world, Pluto leaves a long list of significant sites to explore.

Life in the Cold Lane

As for the discovery of habitats for life, Antarctica plays a role, too. We have explored its caves and found colonies of tiny microbes that might survive even in the harsh conditions of other worlds. Mount Erebus spelunkers like Aaron Curtis foreshadow explorers who will crawl into the caverns on the Moon, Mars, Europa, Titan, and Enceladus. While these alien caves may be forged by forces different from those of Erebus, techniques for exploring them will share much in common. Perhaps, as astrobiologists hope, Erebus' microbial colonies will find commonality on other worlds as well.

Fig. 8.5. Pluto's Sputnik Planitia, a vast ice field of frozen nitrogen, spreads across plains below rugged mountains made of water ice (Painting by Michael Carroll)

Antarctica's sea ice offers other exotic life forms. Psychrophiles inhabit the coldest places on Earth, and some thrive in the solid ices floating on the Antarctic sea. Others survive conditions of bitter cold, darkness, and near-zero oxygen levels at the bottom of the harsh continent's harshest lakes (see Chapter 2). Microbiologists have now sequenced the genetic makeup of extremophiles that thrive in these dismal conditions. Two microbes, *Methanogenium frigidum* and *Methanococcoides burtonii*, survive at the bottom of Ace Lake in Antarctica's Vestfold Hills. There, they endure almost no oxygen[3] and average temperatures barely above the freezing point of water. The microbes are methanogens: they produce methane to carry out biological functions. Methanogens are often cited as an example of microbes that could live in conditions found on other worlds such as Titan. This, too, will energize our drive to explore further.

The Future of Antarctica

With the distances and timelines involved, a trip to the outer Solar System will be more expensive than any expedition mounted in the history of human exploration. Even building infrastructure on Mars poses daunting requirements for resources. An international

---

[3] In fact, oxygen is deadly to these organisms.

endeavor will almost certainly be necessary to get people to these distant frontiers in meaningful and long-term ways. Antarctica continues to shine as an example of what can be accomplished when nations combine efforts and resources toward the common goal of exploration.

What about future outposts and camps in Antarctica itself? Designers are drawing plans for major upgrades to support advanced Antarctic research in the future. McMurdo's infrastructure, first put down in the late 1950s, has been developed incrementally, but the National Science Foundation hopes to see dramatic improvements in the years to come.

McMurdo has been likened to a mining town, a decaying shopping mall, and a truck stop. A 2012 blue-ribbon panel, commissioned by the Office of Science and Technology Policy and the NSF, warned that many of McMurdo's 150 structures are in "imminent need of repair and replacement." The report also stated that the "haphazard arrangement" of McMurdo's layout wastes resources as a result of its inefficiency. McMurdo's aging infrastructure was also singled out as a major concern. The failure to upgrade or replace it, the report said, would "simply increase logistics costs until they altogether squeeze out funding for science."

To that end, the US government has commissioned various architectural firms to redesign McMurdo's footprint. Plans call for reducing McMurdo's total buildings from 104 to just 18, conserving power, water, and other resources. New blueprints envisage a large central edifice with a warehouse for food and other supplies. Plans call for the storage area to be partially encircled by a kitchen/galley area, dining hall, general store, office space for residents and visiting researchers, and an upgraded gym. The main building also includes three dormitories that will accommodate up to 900 people. Crary Lab will be incorporated, connecting to the main structure.

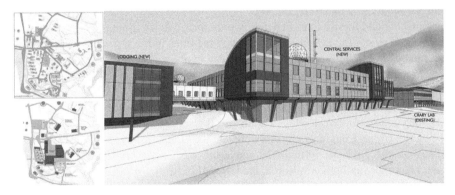

Fig. 8.6. Plans call for dramatic upgrades to McMurdo, including the combining of many small structures into a few larger ones for more efficiency (views at left). At right, an architectural rendering of some of the planned upgrades (Courtesy USAP)

McMurdo 2.0 is a fluid vision, incorporating upgrades to the master plan every two to three years. Several groups are working on the design, including OZ construction, Stevens Institute of Technology, and the Georgia Institute of Technology. The approach to construction encompasses multiple phases, which allows for changes in funding and advances in technology and transportation. During construction, which will take many seasons, no essential functions will be interrupted. Science will carry on as usual, but as the primary base in Antarctica becomes modernized, researchers will be able to work more safely and efficiently.

The United States Antarctic Program is not the only organization with its eye on the future. The British Antarctic Survey (BAS) is also putting into place a stepped program to revamp its flagship Rothera Research Station. The first phase is already under way.

The BAS has established five major research stations across the continent, with additional field huts scattered across the Antarctic wilderness conducting a wide spectrum of science research. The Rothera facility has been operational since 1975. Like McMurdo, its slow development over decades has resulted in an inefficient layout arrangement. Work is underway to combine facilities and rearrange the blueprint to make Rothera a prime example of modern architecture and technology.

The BAS also runs a research station with a creative approach: the Halley VI station is completely portable. The station consists of eight interconnected pods that can be isolated in the event of fire. The complex sits atop hydraulic pillars mounted to 8-meter-long skis. Tractors can relocate the station, which is perched on the sea ice shelf, as needed. In fact, during the 2016/17 season, Halley VI's modules were transported 23 kilometers to a site more stable than the one it had been on since 2012. Many designers foresee this modular approach as the future of facilities at a variety of Antarctic sites.

New Zealand's Scott Base is also advancing toward a more modern infrastructure. The New Zealand government has undertaken a three-year, $6.2 million upgrade. Scott Base recently completed an overhaul of its Hillary Field Center, adding three new laboratories, a mobile container processing and docking facility, workstations for an additional 15 researchers, three meeting rooms, and a common "breakout" center for "cross fertilization within the Antarctic science community."

Both China and the Russian Federation continue to expand upon already established bases. Russia, under the Soviet Union, has had facilities in place since the 1950s. Russia is erecting several stations to support its own version of the US Global Positioning Satellite system. Significantly, Russia was the only country to oppose the establishment of fishing sanctuaries in Antarctic waters.

China, too, has its eye on the natural resources of Antarctica. In addition to fishing rights, the country is interested in Antarctica's oil reserves, which may be the third largest in the world. Chinese engineers recently erected the "Lantern," its fourth research outpost on the continent. The base, formally known as the Taishan, covers 1,000 square meters. It lies near facilities run by South Korea, the US, and Italy.

Many permanent facilities have adopted environmentally healthy designs. One of the most technologically advanced stations on the southern continent belongs to Belgium. The Princess Elisabeth Station is a futuristic-looking pod perched on steel pylons. It is the first Antarctic facility with zero emissions. Established in 2009, the base runs exclusively on wind and solar energy. Its heating is passive; interior temperatures remain at a comfortable level with the use of waste heat from its electrical systems and the warmth from human activity. Its thickly insulated walls allow almost zero heat loss.

Of 13 research bases operated by Argentina, the Carlini is the largest, hosting up to 60 research and support personnel. Carlini has been continually upgraded. Its Dallmann laboratory is a German/Argentine joint endeavor, demonstrating the importance of international cooperation in Antarctica.

International groups have been pooling resources from the beginning of Antarctic exploration. Today, with ever-increasing traffic and expansion of exploration, joining resources has become more important than ever. With its international treaties and its practical need for collaboration, Antarctica stands today as what many hope will be a model of peaceful work in future endeavors.

Farewells

Our last day in the field fell on Christmas day. Everyone present at Lower Erebus Hut, from camp managers to researchers, took the morning to ascend the mountain (one camp manager had to remain at the Hut for safety reasons). It was a sunny, warm day. Some wore light jackets and hats, even in the high altitude chill, while others sported Santa hats. The crater is a remarkable, sheer arena dropping to its active lava lake nearly 100 meters below the rim. Ice coats the cliff sides, outlining cracks and ledges. Sulfur curls up over the crater lip, shading the stony surface in hues of yellow and orange. Perhaps in the shadow of Erebus we were seeing echoes of ancient Mars, when frost may have caked the calderas of Tharsis and Elysium. Perhaps we were witnessing ice towers whose cousins rise from methane shores on Titan, or from the cantaloupe plains of Triton. One thing is certain: we had the privilege of seeing some of the most exotic, beautiful scenes our own world has to offer. One day, planetary travelers will see firsthand what nature has to offer on more distant cosmic shores.

Fig. 8.7. Sunlight casts the faint shadow of an explorer across the mist rising from Mount Erebus' crater. Upraised arms hint at the joy of exploration and the celebration of summiting the remote volcano (Photo by the authors)

9

Journal Entries

Prior to Flight Out

Rosaly

I know plenty of scientists who have been to Antarctica and made some discreet enquiries. My friend and colleague Dr Jani Radebaugh has been on many volcano research trips with me, from Ethiopia to Vanuatu. She has also been on a couple of meteorite collecting trips in Antarctica. Meteorites fall all over the world but they are hard to find in a forest or field. On ice, they stand out, their dark colors easily spotted against ice. Nearly every year a merry, hardy group of scientists heads to Antarctica on these expeditions. Using snowmobiles, they travel long distances, camping out for weeks at a time, returning with a treasure trove of specimens. Jani gave me advice on everything from socks to eating chocolate (if you are really cold, a piece of chocolate will warm you up noticeably quickly). Antarctica is one of the few places in the world where we can eat chocolate without feeling guilty.

In reality, our Antarctic expeditions are relatively easy these days, thanks to communication, technology, and over a century of acquired knowledge. It is impossible to be there without thinking of Scott, Shakelton, Amundsson, and all those brave souls who went to the ice knowing their return was unlikely. We have much to thank them for. However, complacency on the ice can be as fatal now as in their days. The dangerous crevasses are still lurking there, along with many other hazards.

Notes on Packing and McMurdo Station

Mike

> Leave US Dec 11/ arrive Christchurch Dec 13
> Dec 14 get clothing/gear
> Dec 15 arrival in Antarctica ("Ice date") Trainings: field safety, camping, flight, etc.

© Springer International Publishing AG, part of Springer Nature 2019
M. Carroll, R. Lopes, *Antarctica: Earth's Own Ice World*, Springer Praxis Books,
https://doi.org/10.1007/978-3-319-74624-1_9

Dec 22-23 earliest we could leave
Christmas celebration = 26th, no helos
New Year's day, no helos
Try to leave Erebus by Jan 5 (Tues)
Leave McMurdo by Jan 8
On the way back, arrive Christchurch 2 am; first flight 6ish
Arrive US Jan 10
REMEMBER: NOBODY LEAVES MCMURDO WHEN THEY THINK THEY WILL
2-3 days turning equipment in
Getting off Erebus is easier than getting back to New Zealand

Give a couple days advanced notice for helicopter to come

If weather is too good, runway melts

Bring TONS of spare batteries (cold kills them)
Mechanical pencils and sharpies (NO ballpoint pens – they freeze)

Don't forget converter for NZ current

BFC is where you get field camp equipment; they have locked cages that you
can leave stuff in.

There's a small store at McMurdo (like 7-eleven)

July 16, 2016

Rosaly

Got the forms to fill out. If I pass the medical, I'll know I'm really healthy. Fair enough
that they had to have you checked for heart problems and other serious conditions but, wait
a minute, an HIV test? What do they think we'll be doing there? And a syphilis test??? I
didn't know anybody got that anymore. Guess people get bored during the long Antarctica
winter.

Need full mouth X-rays. I suppose you don't want to get a toothache in Antarctica. This
all made me think about the astronaut program. I wonder what they go through. The space
station does not have a dentist yet. I do wonder if they ask for the syphilis and HIV tests.

Partial Packing List (Suggested by Jani Radebaugh and Chris McKay)

Mike

light fleece top
down or fake down med-weight jacket with hood

wind layer on top and bottom – like windbreaker material. Good for keeping
out wind and for being in the tent later and not spilling on yourself...

liner socks – long, very thin socks, you may be able to search under "liner socks" in backountry.com (and there for lots of the other gear)

Thick wool socks – they used to issue those, I may ask Jim if he knows where we can get them, because they are VERY thick (or if you get a chance you can ask him).

White bunny boots – they issue

Snow bibs on bottom – they issue

Little Red – they issue – I like Little Red, but most people wear Big Red, they will also issue that one.

neck gaiter

neoprene snowmobiler mask (look at hunting stores or Sportsmans Warehouse)

hat with ear flaps

goggles - they will issue, I THINK – look for high light blockage, like 50%

glove liners (not cotton) – also look at Backcountry

Thick fake down (real down sweats a lot) ski gloves – look at backcountry. Don't rely on their issue!

October 4, 2016

Rosaly

Email from Mike C: "I'm looking forward to spending some time over cups of coffee with you, free from a screen between us! Of course, there will be howling wind and subzero temps (but hopefully not on the plane)..."

Oh, dear, that does not ease my anxiety...I've heard about those howling winds up on Erebus, so bad that you can't even get out of your tent...can one make coffee under these conditions?

October 5, 2016

Rosaly

The SIP is all approved by NSF and I shopped for more gear while in Syracuse this past weekend while visiting my son Tommy. Found a balaclava which will make me look like a bank robber, but the package did say it was the "ultimate," warmer than any other type. Glove liners and more thermal underwear went in the shopping basket. I don't know what I'll do with all those thermal undies once I'm back. Can't exactly sell them on eBay. Maybe one can but I think it's kind of undignified to sell one's underwear. I can't imagine needing them again as I feel sure that, at some stage during this trip, maybe somewhere on the top of Erebus, I'm going to get desperate from the cold, fall to the ground and raise my fist like Scarlett O'Hara and say "As God is my witness, I'll never be cold again."

October 5, 2016

Rosaly

Amazing coincidence today! My former postdoc, Karl, introduced me to a new postdoc at JPL, also a volcanologist, Aaron Curtis. Young Aaron and I had only exchanged a few sentences when he told me he was going to Erebus in December. "So am I!" I exclaimed happily. We had just been introduced but I felt that having a JPL colleague there would be incredibly reassuring. When Aaron said it would be his 7th trip there, I immediately asked if we could have lunch the next day, which he happily agreed. I mentioned how I was anxious about the cold, as I'm Brazilian. Somehow, I started talking about the thermal underwear I was buying, only later realizing that I shouldn't be talking about underwear (even thermal) with a young man I'd just met. Hopefully he won't tell HR.

Even more amusing is that his former thesis advisor, Phil Kyle, aka "Mr Erebus" will also be there. Phil is something of a legend, perhaps in a notorious way. There are plenty of stories about him displaying behavior that is, let's say, colorful. Frankly, I've long wanted to meet the man, as I like interesting people.

October 13, 2016

Rosaly

Feeling a lot braver after lunch with Aaron, after learning that it's not THAT cold up there on Erebus, usually. Also, the Fang glacier camp, where we acclimate, is only a few miles or so from the Lower Erebus Hut. We won't be so alone and isolated. "It's a good chance to do nothing" he said. I packed up pencils, a space pen (regular pens don't work there), all the thermal underwear. Any day now NSF will get our airline tickets. Wow.

November 13, 2016

Rosaly

I have all the gear, yesterday I went to the camera store to buy bits and pieces, including a camera bag that hopefully I can attach on my belt under the Big Red coat. Batteries drain rather quickly in the cold. I have the video camera for the Erebus lava lake, and my son Tommy's good SLR camera for taking pictures.

I was very excited a week ago when the travel office called me to arrange my flight but don't have a ticket yet. I was just thinking about calling them tomorrow (could an email have gotten lost?) when I opened Facebook this morning and saw the news about the large earthquake in New Zealand. The news was posted by Mike, whose daughter is in Christchurch. He had not heard from her. I can only imagine what that is like for a parent. Luckily later he emailed me saying he heard from her and she is fine. We don't know if this is going to impact our trip. I hope that in the next 4 weeks things go back to normal. The people of New Zealand were lucky, only two deaths reported despite the earthquake being a big one (magnitude 7.5), with an accompanying small tsunami. There are fears of after-shocks and other tsunamis, Christchurch was evacuated. This could throw a big spanner in the works for us. Photos show a lot of destruction. I wonder what the airport is like. I can't find any information on its status.

November 1, 2016: Typical Communiqué to the UTMB

Mike

UTMB Health Center for Polar Medical Operations
  Levin Hall, 5th Floor, Suite 5.527, Route 1004
  301 University Blvd.
  Galveston, TX 77555-1004

  FAX: (409) XXX-XXXX

Included in this package, please find my Form 1428, "Applicant Statement and Release of Liability." It is signed and notarized, per your request. I have been advised that I do not need to submit Form 1429, as I am self-employed. Please let me know if this is not correct.

My endocrinologist has already sent a letter on my behalf, and my primary physician's letter will be sent on Friday or Monday (November 4 or 7).

Thank you!

Michael Carroll (PI)

November 2016, just before Thanksgiving

Rosaly

Michael finally got PQ'ed! Wow, that is such a relief. Now I can be a little more open telling people about the trip. I'm always afraid to say much when there is a chance it won't happen.

I started packing a week before. Made an even more detailed list than my usual packing list. You cannot get enough advice. I was glad that Jani Radebaugh, veteran of meteorite hunting, was always ready to answer my questions. I met her ice buddy Barbara Cohen during a trip to Marshall Space Flight Center, Barbara was also a great source of information.

December 9, 2016

Rosaly

The big day is finally here. My flight is late at night, but I leave the house at 7 pm to give myself a full 3 hours at the airport. Got to work earlier in the day. The hardest part of the trip preparation was to buy a phone card that I will be able to use in Antarctica. Most phone cards these days are for mobile air time. People just can't imagine a place with no cell towers. I wonder how long it will be before McMurdo gets cell service!

The flight was uncomfortable. I like to stay hydrated on planes, which means getting up often to use the facilities, but I was on a window seat. Next to me were an Asian mother and daughter team who seemed to sleep soundly, the teen embracing a giant Mickey Mouse. They were apparently part of a tour to Disneyland.

December 13, 2016

Mike

Christchurch is a beautiful city struggling toward former glory. Its parks, especially along the river, are elegant and idyllic. Statues remind us of historic figures, as do some of the older buildings downtown. But vast swatches of the business section have been razed by 2008's earthquake, replaced by temporary trailers. The cathedral itself looks like a bombed-out building in WWII London. A cavernous stone structure hints at soaring arches and flying buttresses, but half of the gothic edifice is simply gone, removed by the earthquake of 2008. It is a sad testament to two warring factions: those who want to refurbish what is left (preserving its archeological value), and those who want to clear the old structure away and replace it with a new, earthquake-resistant one. It is the symbol of the city of Christchurch; I hope they can bring it back soon.

Christchurch, December 13, 2016

Rosaly

And talking of chickens…I'm still getting used to the kiwi accent. It was a bit of a struggle to understand the instructions yesterday, even though everyone was so friendly. Both Mike and I kept hearing our jolly briefer, Nathan, talk about "chicken" yesterday – eventually we figured out he meant "check in." We are now an hour and a half into the flight, so hopefully I can soon start to count my chickens. After the two-hour mark, there is no turning back, as they won't have enough fuel, so we must make it to McMurdo whatever the weather.

December 13, 2016 (again): Antarctic Center

Mike

Due to the time change, we missed a complete day on the way down. Adjacent to the Christchurch International Airport is a big blue warehouse with a sign reading "United States Antarctic Program, Where Discoveries Begin/National Science Foundation." This is where the clothing distribution center is, along with the computer center, where we will get antivirus software to make our computers ready for McMurdo. Across from the computer and med center, there's a museum with live penguins, a replica of Scott's cabin, gift shop, and restaurant. It's called the International Antarctic Center. They have real snow, rides on a Haaglungs tracked arctic vehicle (apparently we will see them at McMurdo) and very nice theater. One room is a simulated Antarctic storm. Rosaly wanted to go in, but I figured we'd be seeing enough highly realistic storms on the ice. We finally passed it up in favor of completing our computer training.

Required Extreme Cold Weather gear (to be worn or carried on all flights) includes:

Knit hat or balaclava
Goggles
Parka
Mittens/gloves
Wind pants or Carhartt overalls
Boots
Base layers (long underwear/warm socks)

December 13, 2016: Flight to McMurdo

Rosaly

On board the LC-130. Wow, it is rare for Antarctica-bound people to make their ice date. My friend Jani said she never did. I can't believe how incredibly lucky I am. Weather looks beautiful. However, we are not 2 hours into the flight yet, so I shouldn't count my chickens. Two hours is the point of no return. The pilots will check with McMurdo and if the weather looks like it's turning bad there, that is when we can turn around, or get "boomeranged" as they say here. They even have us pack and label a boomerang bag. If we have to turn back, they will return us that bag, while our checked bags will stay on the pallet.

I'm less anxious now than I was even yesterday. There is a lot of information to absorb, but it starts making more and more sense. Yesterday was our clothing day. We watched several movies about Antarctica, all about safety, conservation, and what to do on the day of the flight etc. Of course, they tell you everything that can go bad. The packing is a little daunting, at first it looks like the orange bags they give you don't hold much, but they do. We can take a total of 85 lbs as the combined weight of our checked in, carry on, and boomerang bags. I left my big roller suitcase at the clothing distribution center. It's all really well thought out. I took my suitcase there yesterday with things I knew I'd leave behind. They keep it for you overnight in the changing room and don't put it away until they know you are on the ice. So, this morning I was able to put additional things in that bag. I needed an umbrella yesterday in town but won't need it on the ice!

The Hercules plane has very few windows and they are at high level, so I can't see what is down there, though for most of the journey is only ocean. They give you all these safety instructions about oxygen hoods, life vests, etc. Scary thought. There seems to be nowhere to land before we reach McMurdo.

Mike and I got asked to give a talk to the scientists and staff while at McMurdo, so are combining our talks while on the plane to make for a smoother transition. I call the combined talk a Fred Astaire-Ginger Rodgers act, so after that Mike and I started calling each other Fred and Ginger. Giving a talk at McMurdo is exciting in many ways, including that I'll be able to say I've given public lectures on every continent! Assuming I make it to Erebus, that will mean I'll also have been to active volcanoes on every continent! OK, bragging is not nice, but on these days of Facebook, it seems we do it more and more. But the thought of these milestones really makes me go "wow."

We passed the two hour mark, no turnaround. Wow, we are so lucky. Hopefully we will have great weather on Erebus as well. And on Fang. That is still the main apprehension, 48 hours camping on a glacier. I hate camping and I hate cold. And, strangely enough, so does Michael. It is funny what you learn about someone else on a trip like this. An active volcano is the only lure to make me camp.

Our gear is amazing. The bunny boots are bulky but comfortable. The Big Red coat is a bit of a chore to put on. There are inner gloves and outer gloves or mittens. Hat, ski mask, goggles. I have thermal underwear on, with pants over that, the ski pants over that. I wear three tops: thermal underwear, a fleece jacket and the Big Red.

The flight is surprisingly smooth and comfortable. There has been zero turbulence so far. They let us move around and we have more leg room than in most commercial flights. The Big Red coat makes a cushy seat. By sheer luck, I ended up with the warmest seat on the plane, the woman next to me even complained it was too hot. The soldier "steward" said that we were seated where the hot air blows, so in order to be comfortable elsewhere, it was a bit too warm for us. Trust a Brazilian to end up directly in the path of the hot air!

The Flight Down

Mike

Aircraft exterior description:

Ours was a 4-propeller, conventional C-130 Hercules, specially equipped with skis. The skis add an "L" to the front end of the designation (LC-130). Several of the aircraft have now been fitted with new engines called Scimitars. The props look like black, curved Ottoman sabers. They're more efficient for ski landings, and more powerful for takeoffs without JATOs. Their only drawback is they tend to use more fuel. Our craft had conventional propellers.

In this plane, I feel like I'm in the cast of one of those old WWII movies about the courageous crews in a B-17, braving high-altitude cold in what's basically a flying warehouse. The ceiling and every wall are festooned with cables, copper tubing, and metal shelves holding strange equipment with fat red dials and silver switches. First aid kits swing back and forth in a row above the windows. Mirroring them just below are the life vests. Behind our seats are oxygen hoods in little bags, in case of fire or decompression, which can happen even at this relatively low altitude. It's too loud to talk; they issued us earplugs for a good reason. Instead, Rosaly and I communicate by typing on our laptop screens.

5-gallon water jugs hang from a central structure that houses a row of unused seats. Heat blows from two big pipes on the ceiling. Five hours into the flight, it begins to get really chilly. I guess we haven't seen anything yet.

Our stewards pass out generous lunches: two sandwiches, a candy bar, granola bar, cookie, 2 bags of chips, and fresh apple. They are gearing us up for the high-calorie intake required in a cold and harsh environment.

The cargo master has piled our carry-ons down the center of the cabin. Toward the back of the plane, just where our plane-length canvas benches end, pallets tower up to within inches of the ceiling. Small, round blue lights illuminate our surroundings. We rest our backs against red netting. They warn us to not leave our luggage on the floor if it contains anything that will freeze. Most backpacks hang from the walls on fasteners.

The small portholes reveal nothing unless you stand up and peer through one. Then, unrelenting clouds just below us. Through occasional breaks we can see the metallic-blue sea undulating in great, sluggish waves. We watch the clock anxiously. When we pass the point of no return, an imaginary spot some 2/3 of the way down to the continent, we will know we're committed to pushing on to McMurdo. After a long vigilance, Rosaly taps her keyboard, and I type back:

R: We are definitely past the point of no return. Amazing!
M: BOY is it overcast down there!
R: I was just thinking the same. Hope it's sunny in McMurdo….
M: Whatever way, we are committed to landing now.
R: The trash guy said 2 more hours, thereabouts. I know it depends on winds.

Soon, we begin to see the brilliant white of icebergs, like dustings of powdered sugar on the steely water's surface. Farther south, pancakes of sea ice blossom on the surface like giant water lilies.

In fact, I was initially disappointed that we were taking the LC130 Hercules rather than a jet. Flight times for the big jets are about 5 hrs, ours would be closer to 8. What I forgot to take into account was that our slower aircraft would fly low and slow over the gleaming Antarctic landscape upon our approach. We have 45 minutes of glorious views out the tiny portholes of glaciers, icebergs floating in roundels of sea ice, and rugged black mountains rearing out of the brilliant white ice.

Tuesday, December 13, 2016: Arrival on the Ice

Mike

Because of the lack of windows, the landing is a bit mysterious. We must guess what is going on by listening. We can also see – up against the ceiling – twisting tubes leading to the flaps as the pilots raise and lower them. The landing is gentle, but we hear the scrape of the snow and ice on the landing skids.

The veterans give us sideways glances as we bundle up in our ECW gear. A pilot sitting next to me works for Kenn Borek Aviation, the heroic Canadian air service that ties Antarctic remote camps together. He is wearing a lightweight flight jacket, sunglasses, and no hat. I have on my Big Red Parka, hat, gloves, Baffin boots, Carhart bib overalls, and several toasty base layers. The hatch opens at the back, and our steward pops open the front doorway. Cold air blasts into the cabin, along with blazing white sunlight. I am glad I'm bundled up, but once outside, we acclimate to the sunny, brisk environs.

We step out and see Erebus towering behind the aircraft. McMurdo is essentially in the foothills of the big volcano. We drop off some folks at the New Zealand station, Scott Base, which is just down the hill from McMurdo. McMurdo resembles many of the studies I've seen that project what a Mars settlement would look like. Although heavily organized and very neat (you see no trash anywhere), the place has a distinctly industrial feel. The village swells to over a thousand in the summer season. It has its own hospital, fire department, library, two bars and a quiet coffee house for the more contemplative. It is a spectacular bastion of civilization in a remote and hostile place (kudos to NSF and USAP).

Wednesday, December 14, 2016

Mike

Elaine tells us that McMurdo has three seasons: dirt, mud and ice. We are in the dirt season. They say it is much prettier in the snow. Temperatures are comfortable, in the way they are in a mountain town like Estes Park on a sunny winter day. The sun is out and there is no wind. But because this place is surrounded by ice (sea ice, mountain glaciers, frozen plains) the least breeze drops the temperature by a dozen degrees or more. I can see why wind is a serious concern here.

McMurdo is like a college campus with no grass (we hear there is a greenhouse, but we haven't seen it yet). The rooms are arranged like a college dorm. Common areas are on the ground floor. Each room has two beds, two desks, and a heavy blind to keep out the 24/7 daylight for sleep periods. The literature says, "Due to numerous work schedule differences, there will always be people sleeping in the dorms at different times of the day and night. Everyone is entitled to his or her rest, so please be considerate of your fellow workers and observe 24-hours/day quiet hours in the dorms." People were very cognizant of this, so the hallways felt like a library or sanctuary, very quiet.

The floor's laundry room was just across the hall from my room, and the restrooms were down the corridor a short way. The restroom had the feel of a locker room with showers. Laundry soap and washer/dryers are free (as we have paid for them through our taxes), which is wonderful. No searching for quarters.

McMurdo's medical clinic, bldg. 142, is next to the big blue building that houses the store, galley, offices and only ATMs on the continent. It's an impressive facility, with several beds, examination rooms, pharmacy, X-ray, and a gorgeous quilt of a penguin hanging behind the main desk. Several RNs or Pas are there during posted hours, and available by phone 24/7. My goal is to visit there only as a tourist.

Our introductory lit gives tips for maintaining health:

Drink more water than usual to avoid dehydration.
Dress in layers. Use toe and hand warmers (provided at work centers or Central Supply)
Use snow traction devices on boots/shoes
Use sunscreen (provided at handwashing stations) and 100% rated UV sunglasses
Exercise regularly; maintain healthy diet.
Wash hands often.
Get plenty of sleep.

They make the "healthy diet" part easy with the great food available at the galley. The galley is open 24/7, and like the laundry facilities, the food is free. The "lots of sleep" part was something I couldn't follow. There is a whole lot of adjusting going on here, from time zones to weird light to the stress of just keeping the schedules straight.

Each room has a TV with closed circuit channels. These are somehow related to the armed services broadcast network. The weather channel has a report: Yesterday's high at McMurdo was 34°F, with a low of 16°F. Next sunset will be at 1:23 am local time…on February 20! Tonight's forecast: sunny.

Amongst the hodge-podge of buildings, some tattered and old, others fairly modern, is a little white building that sticks out. It looks like a miniature church, standing at the precipice between McMurdo's campus and the McMurdo Sound below. The sea behind is frozen, and the distant Trans-Antarctic Mountains like Mount Discovery rear up in ragged masses of black rock, framing the little building. The place is called "Chapel of the Snows" and has a colorful history. McMurdo's original plan did not call for any specific building dedicated to the spiritual side of life. But during initial construction in 1956, a "mysterious pile of lumber, planks, nails, Quonset hut sections, and assorted materials began to accumulate on a knoll overlooking the camp," according to the Chapel's brochure. Volunteers took the leftover materials and built the non-denominational chapel in their own spare time. The popular chapel was destroyed by fire in 1978 (rescuers managed to save the bell and stained glass), but was so popular that two other successive chapels were fabricated. Today's Chapel of the Snows was also built by volunteers with material salvaged from older structures around McMurdo. It's a nice place to go for meditation, reading, prayer, or just admiring the Islamic, Jewish and Christian objects inside. The day was sunny, so between classes I sat outside and did an on-site painting of the Chapel and mountains behind. The afternoon sunlight gives the mountains the appearance of molten glass.

Our Upcoming Schedule

Mike

Our stay will be a week of intensive training and tours. This is the list they gave us in preparation for our classes. Our Field Support and Training Scheduler, Abby Vogler, is available and helpful, which seems to be a theme around here! She listed several courses we would be enrolled in that are specific to our project:

1. Altitude Course: Wednesday, December 14 at 8:30 am

This 2-hour course is designed to introduce altitude physiology, illness, prevention, and treatment information. The instructor will also demonstrate practical usage of a Gamow bag. This is required if you will be working above 8,000 feet without close support.

2. Antarctic Field Safety: Wednesday December 14 at 1:00 pm

The primary USAP Field Safety course, required for all USAP participants traveling to a field setting.

Afternoon schedule:
1:00 – 3:45 Antarctic survival skills, personal wellbeing, and risk-management theory
3:45-4:15 Helicopter info

3. Glacier Travel/Crevasse Rescue: Thursday December 15 at 9:00 am

This single-day course is required if you will be traveling in a crevassed area. You will be learning and practicing basic glacier travel techniques: ice axe usage; self-arrest techniques; and crevasse rescue, roping and harnessing.

We were also given a general training checklist.

<u>**Training Checklist**</u>

Instructions [digital]: *Click on any trainings that are required for this group/event. Be sure to save your changes before closing.*

Required Meetings and Trainings

☐ Science Inbrief
☐ CORE training (Classroom Vehicle training, Fire Safety, Waste Briefing, Medical Briefing)
☒ Practical Vehicle training
☒ Snowmobile training (standard)
☐ Snowmobile training (advanced)
☒ Crary Lab Safety Orientation
☐ Meeting with Crary Lab IT to gain access to wireless network
☒ Environmental Field Brief (Required for all McMurdo grantees)
☐ McMurdo Dry Valley ASMA Environmental Training (Required for all grantees going to the McMurdo Dry Valley ASMA)
☐ FST Training Classes – please see the RSP for specific classes.
☒ MacOps Pre-Field Communications Briefing (for any group going to a field location)
☐ Field Food Pull

Note Food is not available for day trips from the BFC, please use the galley grab and go.

Thursday, December 15, 2016

Mike

Our training for crevasse survival was exciting. Valentine Kass, director of the Artists and Writers program, came with us to see—and take part in—the training. She is fearless, always willing to volunteer to climb into a Gamow bag or belay into an abyss.

We finally got out into the brilliant blaze of Antarctic sunlight. Our Hagglunds vehicle, driven by mountaineer Evan Miller, ascended the hills outside of McMurdo on a crushed-lava road, passing various structures, the Fire House, and fuel depot. From there, we dropped down toward the coastline, passing the bright green buildings of Scott Base. Before heading out into the true wilderness, we pulled over to a parking area, where we used stiff brushes to brush off the vehicle. Then, we drove over a metal grate to knock off any dirt from inhabited areas. They go to great lengths here to protect Antarctica's environment.

Evan positioned us at the base of a slope that had several hidden crevasses. Snow builds thin bridges over these deep cracks, making them treacherous. Black flags warn travelers of danger, and a long train of green ones leads us up to the danger spot. But Evan decided that it was too dangerous for us to use this crevasse, so we went down to a gigantic trench dug for just such training.

The trench was actually over the ocean, dug into the Ross Sea Ice. We rescued a heavy backpack from doom, and then I climbed down into the ditch. I could see a beautiful progression of blue, tan and grey layers, a record of eons on the Ross Ice Shelf.

December 18, 2016

Mike

Tomorrow, we fly to Fang Glacier, and then on to the Lower Erebus Hut two days later. We have to be super organized. If we forget anything, we do without. This is going into the deep wilderness, and it's a little intimidating. Main concern is not me freezing, but freezing insulin. Will keep several stashes separate and one on me.

Bag drag is tonight. Here's my list:

12/19
Michael Carroll helo baggage
Building 203A Room number 214
Event # W487

Coat
Insulin
Cold weather head stuff
Gloves
Insulin (both kinds)
Batteries (both sizes)
Orange duffel (mark DNF, but will go in helo, so may freeze)
Paint/canvas
Pump supplies
backup pump
insulin/pens/needles
CGM inserter/charger
Fuzzy hat/blue glove liners/ extra balaclava/extra wool socks
Boston baked beans/kind bar
one base layer pants/2 base layer shirts/undies
long sleeve nite shirt
Christmas present/Rudolph hat
Hiking boots
Wash cloth

Backpack
Radio,   all injectors
Pee bottle,   all oral meds
Book/papers   socks x2
Toiletries   one pair underpants

Toothbrush/paste
Computer   wipes
Tester   SUNBLOCK/hand cream
H2O bottle   vaseline
chicken/TimTams/candy

[WEAR one set base layers]

December 19: Infamous Fang Glacier

Mike

Fang glacier is as majestically intimidating as its name. It rests uncomfortably on the shoulder of Mount Erebus, a flat area with nothing for miles around but four tiny, forlorn Scott Tents, spots of yellow on a vast grey-blue plain. The helicopter flight to the camp is one of the most spectacular flights I've ever been on. Gliding over the great frozen sea, we can spot ice leads, fractures, and a varied palette of umbers, tans, greys, cobalt and even crimsons, all locked just beneath the grey-white powdery surface of sea ice. We float over crevasses as the mountain rears up. The ever-present plume issues from the top of the mountain, but we head to the left and circle around, ending up skimming over a surface of white ice and black rock. The helicopter lands in a blizzard of snow, kicked up by its own hurricane. As it settles, and as the prop comes to a stop, silence sets in. The weather is perfectly calm. The air is thin and cold. We offload our gear and lay atop sleeping bags and crates as the chopper takes off again. It disappears into the cerulean sky, our last physical contact with civilization for a while.

We stow our gear, get in radio contact with McMurdo, and then settle in for the next two days and nights. Evan is a slave driver of inactivity: he keeps us quiet and hydrated. The closest tent to our landing spot is a supply tent filled with food, propane, and various crates and duffle bags. The next is to be home for Rosaly and me. The third tent is Evan's, but doubles as our communal area and cook tent/dining hall. The fourth is the toilet teepee, the poop pavilion. For heat, Evan boils water. But we are at an altitude of 9,000 feet, so the water will not get as warm as it would down at McMurdo. Still, it warms our insides and gives some heat to the tent. Because of the danger of carbon monoxide, the rule is that if we have a camp stove on, the front door stays open. Meal time is chilly when I sit by the door. Evan's meals are hearty and hot. He has a secret weapon to keep us warm at night: he turns our drinking Nalgene flasks into hot water bottles that we keep inside the sleeping bags with us. We grudgingly go back to our tent to try to read in the cold air. Snow covers our tent floor, still frozen, like powdered sugar. There is plenty of light, but we are too cold to read or write. We brought playing cards, but we're too oxygen deprived to think.

The next morning I discover that my wipes are frozen into a solid block, as is the bottom of my drinking water. My glasses are so cold that they stick to the skin on my temples. After breakfast, Evan takes us on a short hike. The landscape before us is desolate, stark, beautiful. We can see the ocean and sea ice 9,000 feet below, down the steep mountain. A bastion of rock rises from the glacier, surrounded by an ice crystal halo. I will never forget this scene.

December 22, 2016: The Ice Caves

Rosaly

On our first day at the Lower Erebus Hut, the weather was not good, but staying put helped us acclimate. I feel terribly sluggish; it is hard to sleep soundly at high altitude, and the rack tent, although luxurious compared to regular tents, lets in a lot of light. Before dinner, Evan thought we could go to the first of two ice caves we were allowed to go in – he had scouted it earlier in the day. It was Mike's (brilliant) idea to get us permission to go into the cave. I was not prepared for the sheer beauty of it. It was breathtaking. I felt transported to a magic world where princesses lived in icy splendor. I wished every girl who loved the movie "Frozen" could see this. Photos don't capture the beauty.

December 22-3, 2016

Mike

Caves are AMAZING! The stony floor is dry, so must be pretty cold, because the chamber is filled with vapor. Every kind of blue imaginable—from the phthalo blue of glacial ice to the near-purple of ultramarine—and textures galore. The crystals form in the breeze. They indicate the direction of air flow, but oddly, they grow toward the flow, not along with it. We smell fresh-frozen water, sulfur, and the plastic/canvas of our gear. The lens of the SLR fogs rapidly, but I get a few images. Evan seems as fascinated as we are. I'm glad our trip has not made him bored; he's an experienced mountaineer, but so patient with us, and happy to share the joy of discovery. Light filters through the ceiling in little squiggles that delineate fractures in the ice, and you can see where one might fall through from above, but ominously, on top—in the daylight—you can't tell where the chambers are at all. No difference in the ice. That's why Evan is so careful here, prodding the ground in front of us with his crevasse pole.

The towers above the caves are in constant change, sculpted by wind and growing by the varying flow of hot air from the heart of Erebus. Towers build in blades, turrets, corkscrews. Reminds me of some of the formations in the Devil's Golf Course in Death Valley. Some are sixty feet tall, easily.

December 24, 2016

Rosaly

"Christmas Eve is when Brazilians have their big celebration and exchange gifts. It feels strange to be so far from family, but luckily we can keep in touch via email."

December 24, 2016

Mike

Christmas Eve at the Lower Erebus Hut. The camp managers are preparing a feast for tomorrow, Antarctic style. LEH kitchen and common room are one and the same. The "freezer" is the outer mudroom between the inner and outer doors. The "refrigerator" is the shelf against the wall under the sink. Drinking water is melted from the snow outside. Once in a while, one gets a whiff of sulfur from the volcano's breath, but rarely. Kitchen cleanup is a shared duty, as is cooking. Scientists coming in from the field share one thing in common: exhaustion. But spirits are high and the essence of Erebus invigorates everyone. We still feel the excitement of the caves we have seen, the pillars of ice we've explored, the travels across the precipitous glacial ices leading to Tower Ridge. A row of Santa hats hangs from a clothesline. One of the camp managers wears fuzzy reindeer antlers, while the other keeps the top of her head permanently obscured by a knit cap.

   Christmas Eve. The last one I had was at home, opening some presents and having a family dinner with kids, sister, aunts and uncles, grandparents and friends. My Mom and Dad are gone now, but I know they would be happy to know I was spending the evening on the side of a steaming volcano with a bunch of happy scientists and engineers.

December 25, 2016: Christmas Day

Mike

We have had a wonderful, creative feast. The clever camp managers of the LEH have treated us to a truly fine Christmas dinner. Now, everyone but one person (who must stay behind for safety reasons), ventures to the crater rim of the great Mount Erebus. The drop is vertiginous. Down there, some five hundred feet below, shrouded in sulfurous mist, lies the lava lake Rosaly would so love to see. Our trip to the rim is icing on the cake, and even if the view was obscured, just being here, gazing down the throat of an Antarctic volcano, is a privilege. As I look down on this very special and spiritual day, I can see Evan's shadow in the billowing steam. His arms are outstretched in a familiar form. I can't help but be reminded of a carpenter from Nazareth, for whom this day is celebrated by so many. But in this silhouette, I also see the pure joy an explorer feels after a hard-earned trip to a pinnacle, a place where the world seems new and fresh.

Glossary

Acute Mountain Sickness (AMS)   An illness that can affect hikers, skiers, or other travelers at high altitudes, usually above 2,400 meters. Symptoms may include dizziness, headache, shortness of breath, and rapid heartbeat.

ANSMET   Antarctic Search for Meteorites.

ASPA   Antarctic Specially Protected Area. These sites are designated to preserve historically significant or biologically sensitive contents.

benthic   Zone at the bottom of a body of water, such as the sea floor beneath the Ross Ice Shelf.

Big Red   Official USAP down-filled parka, issued to all scientists and staff serving at McMurdo, Palmer Station, and South Pole Station. The coat is bright red in color.

boomerang flight   A flight that turns back to its point of origin before reaching its destination, usually due to weather.

Carhart bib   Wind-resistant overalls, part of the ECW set.

Condition One   Antarctica's most severe weather conditions, with winds reaching hurricane force and visibility less than 100 feet.

Condition Three   Fair weather, with warmer and calmer conditions than Condition Two.

Condition Two   Weather more moderate than Condition One, with visibility less than ¼ mile and winds between 48-55 knots.

crevasse   Deep fissure in a glacier or ice field, often covered over by a thin bridge of snow, making it invisible from the surface.

cryolava   Erupted materials, mostly water, that flow out of a cryovolcano.

cryovolcano   A volcano that erupts volatiles like water and ammonia rather than molten rock.

dirigible   Gas-filled airship with a rigid structure. Such vehicles have been used for Arctic and Antarctic exploration, most notably Umberto Nobile's Norge.

dirty cave   Ice cave that has been exposed to unprotected humans and so is accessible without a biosuit.

ECW clothing   Extreme Cold Weather clothing issued or approved by NSF guidelines.

Erebus crystals   Rare feldspar crystals that form in only two volcanoes on Earth, Mt. Kenya in Africa and Mt. Erebus in Antarctica.

extremophiles   Organisms capable of thriving in extreme conditions like bitter cold or high-salt settings.

feldspar   Mineral containing alumina and silica.

FSToP   Field support training personnel.

fuelies   Fuel operators that service many locations in Antarctica, from McMurdo to remote field camps.

fumarole   A volcanic vent that emits steam and gases.

fumarolic cave   Ice cave formed over a fumarole, where volcanic gas (mostly hot water vapor) melts a cavern within the ice overlaying the fumarole.

Gamow bag   Inflatable bag used to treat a person suffering from critical high altitude sickness, usually inflated by a foot pump or electric compressor.

glacier   A river of ice.

guttae   Dark, pond-like features on Neptune's moon Triton.

Hagglunds   Swedish, all-terrain, tracked vehicles used in arctic conditions. They are amphibious and can float in water.

Hercules LC-130   Ski-equipped C-130 transports, currently operated in Antarctica by the 109th wing of the US Air National Guard, used primarily for transport between Christchurch and McMurdo.

High Altitude Cerebral Edema (HACE)   Along with heightened symptoms common to AMS, HACE may manifest as confusion, fever, and altered mental state.

High Altitude Pulmonary Edema (HAPE)   The most serious of high altitude sickness, HAPE may accompanied by difficulty doing simple tasks, shortness of breath at rest, and a cough that produces frothy sputum, sometimes tinged with blood.

ice tower   Ice structure built above fumarolic caves at Mount Erebus. These dynamic structures change with wind, temperature variations, and other weather conditions.

ionosphere   The layer of the earth's atmosphere between 80 and 950 kilometers, containing high concentrations of ions and free electrons. The ionosphere has historically been used to reflect radio waves.

International Geophysical Year (IGY)   Lasting from 1957-1958, IGY was an international effort to coordinate the collection of geophysical and other scientific data from around the world. The year's activities marked an expanded reengagement of the world's scientific community after the Cold War.

Ivan the TerraBus   A terra bus is specially designed to carry passengers across ice and snow. Ivan is the most famous of McMurdo's vehicles.

Jamesway Hut   Insulated quonset hut designed specifically for arctic conditions.

karst   Terrain formed from the dissolution of rocks such as limestone, dolomite, and gypsum, characterized by underground drainage systems with sinkholes and caves.

karstic lakes   Lakes formed by collapse of subterranean caves, especially in water-soluble rocks such as limestone, gypsum, and dolomite.

lava lake   Pond of molten lava contained within a volcanic crater or caldera. Active lava lakes are rare, found only in a few places on Earth today: Erta Ale in Ethiopia,

Nyiragongo in the Democratic Republic of Congo, Kilauea in Hawaii, Mount Erebus in Antarctica, and Ambrym in Vanuatu, Villarrica in Chile.

MacOps   The control center for McMurdo Station's operations, including air traffic control and communications.

magnetosphere   Energy field surrounding a planet or moon, usually generated by a molten core.

Maori   The indigenous Polynesian people of New Zealand.

Milvan   Military Owned Demountable Container used for shipping loads to and from Antarctica.

National Science Foundation (NSF)   United States federal agency established in 1950 that supports and enables research and education in the fields of science and engineering.

paleolake   An ancient lake or remnant of ancient lake.

patera   Irregular, flat crater, usually with scalloped edges, frequently associated with volcanic activity.

phonolite   Rare type of lava rich in nepheline and feldspar.

PI   Principal Investigator, team leader or researcher in charge of a funded NSF project.

polar gigantism   The tendency for invertebrates and other deep-sea dwelling creatures to grow larger in size than their warmer water relatives.

pre-qualification (PQ)   Physical qualification required to travel to Antarctica.

Rothera Research Station   British flagship station in Antarctica, situated on Adelaide Island, operated by the British Antarctica Survey.

Scott tent   Four-sided pyramidal tents based on designs by Robert Scott for his 1910-13 South Pole expedition. They are stable in wind, standing roughly 9 feet tall with an 8-foot-square base. Commonly used in remote camps in Antarctica.

shield volcano   Gently sloping volcano built of thin lava flows. Shield volcanoes display a profile similar to the shield of a warrior and are common in Iceland and Hawaii.

Skua bird   An aggressive brown sea bird that frequents Antarctic shores.

Skua exchange (verb skuaed)   Tradition of recycling unneeded equipment. When personnel depart Antarctica, they often donate usable items to the Skua, to be traded by those left behind. The Skua is a valuable source of items for travelers and inhabitants alike and reflects the importance of recycling in the closed communities of Antarctica.

sloop   Single-masted boat with a fore and aft sail.

SPoT   South Pole Overland Traverse.

stratovolcano   Steep-sided, conical volcano typified by explosive eruptions.

Strombolian   Continual, relatively mild eruptions which sometimes eject lava bombs.

subglacial   Occurring beneath glacial ice.

Terra Australis   Vast southern continent shown on early maps of the southern hemisphere; Terra Australis was a theorized landmass, with no direct observation to confirm it.

University of Texas Medical Branch (UTMB)   Health center and medical school, the UTMB is responsible for pre-qualification of all personnel destined to Antarctica under the auspices of the USAP and NSF.

USAP   United States Antarctic Program. The USAP oversees all US scientific research and logistics in Antarctica, as well as flights and shipping in the Austral (Antarctic) Ocean.

WAIS Divide   West Antarctic Ice Sheet Divide, the high region that separates ice flows bound for the Ross sea from the ice flows headed to the Weddell sea.

Weddell Seal   Leptonychotes weddellii; a large seal with the most southerly distribution of any mammal. Weddell seal range includes the region completely encircling Antarctica.

Assorted Landmarks

Amundsen/Scott Base The outpost at the geographic South Pole.

Christchurch New Zealand's southern city, which serves as staging area for flights to Antarctica (particularly McMurdo).

Crary lab (Albert Crary Science and Engineering Center) A five-pod structure dedicated to laboratory and office space for scientific research by visiting researchers. The lab has dedicated areas for Earth Sciences, Biology and Marine Biology, and Atmospheric Sciences. It also has an aquarium.

Discovery Hut Built on Hut Point Peninsula, the historic wooden structure was built by Robert Falcon Scott for his *Discovery Expedition* (1901–1904). The hut was later used by other expeditions, including Shackleton (1908) and the *British Antarctic Expedition* of 1910–1913.

Enceladus Ice moon of Saturn exhibiting geyser-like activity. Structures related to the Enceladus jets might resemble some formations seen on Mount Erebus.

Europa Ice moon of Jupiter. Superficially, Europa's frozen surface resembles some sea-ice formations and ice fields in Antarctica.

Gallagher's Popular pub at McMurdo Station.

Hut Point Peninsula A long, narrow peninsula from 3 to 5 km wide and 24 km long, extending southwest from the slopes of Mount Erebus on Ross Island, Antarctica. The end of Hut Point is the location of Scott's Discovery Hut, for which it is named.

Io Volcanically active moon of Jupiter; has lava lakes analogous to the one found at Mount Erebus.

Lower Erebus Hut Permanent facility on the edge of the Mount Erebus caldera. The hut is used to stage to various remote sites on the volcano.

McMurdo Station Main base of operations for the USAP, McMurdo shares resources with other nations and provides a staging area for many research teams and travelers.

Mount Erebus Active volcano on Antarctica's Ross Island.

Mount Terror Dormant volcano adjacent to Mount Erebus. Mount Terror has a repeater station that aids in communication between McMurdo and field camps like the Lower Erebus Hut.

© Springer International Publishing AG, part of Springer Nature 2019

M. Carroll, R. Lopes, *Antarctica: Earth's Own Ice World*, Springer Praxis Books,

https://doi.org/10.1007/978-3-319-74624-1

Observation Hill   Known as "Ob Hill" by locals, this 230-meter-tall hill overlooks McMurdo Station and is a favorite climb on clear days by personnel stationed at McMurdo.

Palmer Station   US research facility on Anvers Island along the Antarctic Peninsula. It is the only US base located north of the Antarctic Circle and is primarily dedicated to marine biology.

Punta Arenas   Southern city in Chile's Patagonia region, used as staging site for shipping to Antarctica.

Ross Island   Island consisting of four adjacent volcanoes, situated in the Ross Sea off the coast of Victoria Land in McMurdo Sound. It is the site of Scott Base and McMurdo Station.

Scott Base   New Zealand's flagship Antarctic research station. Scott shares the helipad and some power with McMurdo Station.

Titan   Largest moon of Saturn, a world with possible cryovolcanoes.

Williams Field   3000-meter snow/ice runway on the Ross Ice Shelf. Serves as the major takeoff and landing site for air traffic to McMurdo Station and Scott Base.

Wright Mons   Possible cryovolcano on Pluto.

Suggestions for Further Reading

Antarctica:

Endurance: Shackleton's Incredible Voyage. Alfred Lansing, introduction by Nathaniel
  Philbrick (2015). Basic Books.
Antarctica: Secrets of the Southern Continent. Edited by David McGonigal (2008). Firefly
  Books.
Alone on the Ice: The Greatest Survival Story in the History of Exploration. David Roberts
  (2014). W. W. Norton & Company.
Antarctica: Life on the Ice. Edited by Susan Fox Rogers (2007). Travelers' Tales.
Terra Antarctica: Looking into the Emptiest Continent. William Fox (2005). Trinity
  University Press.

Planetary Science:

Alien Volcanoes. Rosaly Lopes and Michael Carroll (2008). Foreword by Arthur C. Clarke.
  Johns Hopkins University Press.
Alien Seas, Oceans in Space. Edited by Michael Carroll and Rosaly Lopes (2013).
  Foreword by James Cameron. Springer.
Io After Galileo. Edited by R. M.C. Lopes and J.R. Spencer (2006). Springer Praxis.
Planetary Geology: An Introduction (Second Edition) Claudio Vita-Finzi and Dominic
  Fortes (2013). Dunedin Academic Press Ltd.
Planetary Sciences (Updated Edition). Imke de Pater and Jack J. Lissauer (2015).
  Cambridge University Press, UK.
Planetary Surface Processes. H. Jay Melosh (2011). Cambridge University Press, UK.
Volcanic Worlds: Exploring the Solar System's Volcanoes. Edited by Rosaly Lopes and
  Tracy Gregg (2004). Springer Praxis Books.

© Springer International Publishing AG, part of Springer Nature 2019   185
M. Carroll, R. Lopes, *Antarctica: Earth's Own Ice World*, Springer Praxis Books,
https://doi.org/10.1007/978-3-319-74624-1

Space Art:

The Art of Space: The history of space art, from the earliest visions to the graphics of the modern era. Ron Miller (2014). Zenith Press.

Space Art: How to Draw and Paint Planets, Moons, and Landscapes of Alien Worlds. Michael Carroll (2007). Watson-Guptill.

Visions of Space. David A. Hardy (1990). Paper Tiger Press.

Volcanoes:

Encyclopedia of Volcanoes (2nd Edition). Edited by Haraldur Sigurdsson, Bruce Houghton, Steve McNutt, Hazel Rymer, John Stix (2015). Academic Press.

Modeling Volcanic Processes: The Physics and Mathematics of Volcanism. Edited by Sarah A. Fagents, Tracy K. P. Gregg, Rosaly M. C. Lopes (2013). Cambridge University Press, UK.

Volcanoes, a Beginners Guide. Rosaly Lopes (2010). Oneworld Publications, UK.

The Volcano Adventure Guide. Rosaly Lopes (2005). Cambridge University Press, UK (Portuguese translation by Oficina do Texto, 2008).

Volcanoes, Global Perspectives. John P. Lockwood and Richard W. Hazlett (2010). Wiley-Blackwell.

Volcanoes of the World (3rd Edition). Lee Siebert, Tom Simkin, Paul Kimberly (2011). University of California Press.

Volcanoes (2nd Edition). Peter Francis and Cliver Oppenheimer (2003). Oxford University Press.

Fiction:

Europa's Lost Expedition: a scientific novel. Michael Carroll (2016). Springer Publishers
Antarctica. Kim Stanley Robinson (**1999**) Bantam.

Index

© Springer International Publishing AG, part of Springer Nature 2019
M. Carroll, R. Lopes, *Antarctica: Earth's Own Ice World*, Springer Praxis Books,
https://doi.org/10.1007/978-3-319-74624-1